Interpreting Aerial Photographs to Identify Natural Hazards

Charles E. Glass, Ph.D., P.E.
Associate Professor Emeritus, Department of Mining and Geological Engineering, The University of Arizona, Tucson, AZ

ELSEVIER

AMSTERDAM • BOSTON • HEIDELBERG • LONDON
NEW YORK • OXFORD • PARIS • SAN DIEGO
SAN FRANCISCO • SINGAPORE • SYDNEY • TOKYO

Elsevier
225 Wyman Street, Waltham, MA 02451, USA
The Boulevard, Langford Lane, Kidlington, Oxford, OX5 1GB, UK
Radarweg 29, PO Box 211, 1000 AE Amsterdam, The Netherlands

First published 2013

British Library Cataloguing in Publication Data
A catalogue record for this book is available from the British Library

Library of Congress Cataloging-in-Publication Data
A catalog record for this book is available from the Library of Congress

ISBN: 978-0-12-420018-0

This book has been manufactured using Print On Demand technology. Each copy is produced to order and is limited to black ink. The online version of this book will show color figures where appropriate.

Printed and bound by CPI Group (UK) Ltd, Croydon, CR0 4YY

Transferred to digital print 2013

CONTENTS

LIST OF FIGURES

LIST OF TABLES

PART I

Establishing Baselines: *Look for the Butterfly*

I once read a novel, the name of which I can no longer remember, in which a computer specialist was looking out the window on the top floor of a high-rise hotel. He noticed a butterfly fly past the window. What an anomaly! What was a butterfly doing up there? There were no flowers for it—no mates either. This impressed the computer specialist—and obviously me as well—for when pouring over reams of data from then on, the specialist always looked for the butterfly—that small, yet significant, anomaly that just didn't belong. This will be your job when looking for hazardous ground conditions using aerial photographs. *Always look for the butterfly.*

We will see that dangers are revealed on aerial photographs primarily because they depart subtly from the norm; they are anomalous features that depart from what we expect to see in natural terrain. So searching for dangerous ground conditions on aerial photographs is similar to a forensic specialist searching for clues to a crime.

Look for a moment at the aerial photograph shown in Figure I.1. This is our subject. This is what we will spend the rest of our time studying. What you see in Figure I.1—or more precisely what you *think* you see—depends a lot on how much time you've already spent looking at aerial photographs (your experience). Everyone will agree that aerial photographs frequently contain complex patterns—patterns

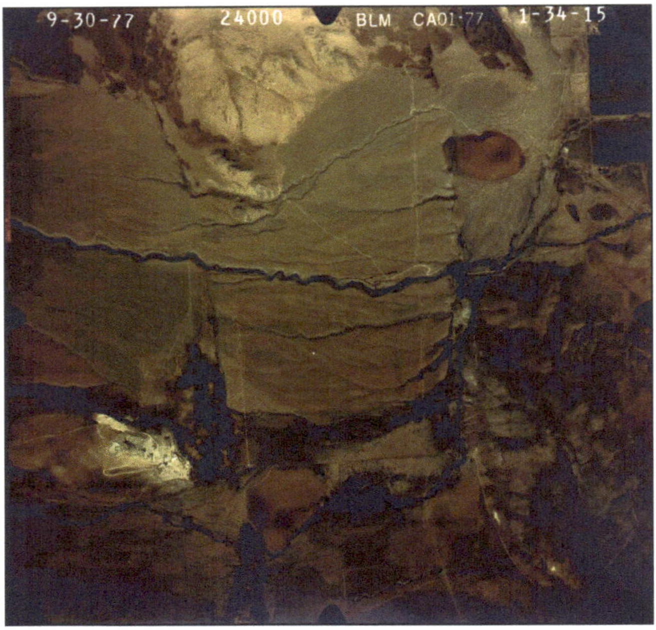

Figure I.1 The color vertical aerial photograph is of Owens Valley, California. North is toward the top. **Place a bookmark on this photograph, or print it, as we will be referring to it later.**

caused by reflection of sunlight from landscape features. We will use these patterns to glean from aerial photographs information about the terrain, geology, vegetation, and land use within the area covered by the photograph. This information will comprise the norm, or the background, from which we will search for butterflies, anomalous patterns that constitute natural hazards. Perhaps you have already seen some natural hazards on Figure I.1. If you haven't, don't be concerned, you soon will.

CHAPTER 1

Getting Started

1.1 INTERPRETATION BEGINS WITH THE SUN

To begin, let us consider the sunlight that we use to illuminate our subject. This will help us understand aerial photographs vis a vis other forms of remotely sensed images.

Sunlight is simply electromagnetic radiation. Electromagnetic radiation comprises a continuous spectrum of energy extending from long-wavelength (low-frequency) radio waves to short-wavelength (high-frequency) gamma and cosmic waves. A cartoon depicting a portion of the electromagnetic spectrum is presented in Figure 1.1. The spectrum is arbitrarily divided into the following regions: x-ray, ultraviolet, visible, infrared, microwave (radar), and radio. Boundaries between these regions are rather arbitrary, and some regions (ultraviolet, visible, and infrared are examples) are further divided into subregions, so you will find a fair amount of disagreement about region boundaries within the literature.

One region of the electromagnetic spectrum that is tightly defined, though, is the visible region, because it is defined by the response of the human eye. Our eyes are sensitive to electromagnetic radiation having wavelengths ranging from approximately 0.4 μm (violet) to approximately 0.7 μm (red).

Through some stroke of luck, evolution, or genius this range of electromagnetic energy also corresponds to the location of the Sun's peak energy output. How fortunate for us; it enables us to see our world. This e-book will deal nearly exclusively with the visible spectrum, even though sensors have been developed that are capable of imaging in other, more exotic, regions. Why are we to be discriminated against and forced to remain in the "low-tech" visible region? There are several good reasons to do so.

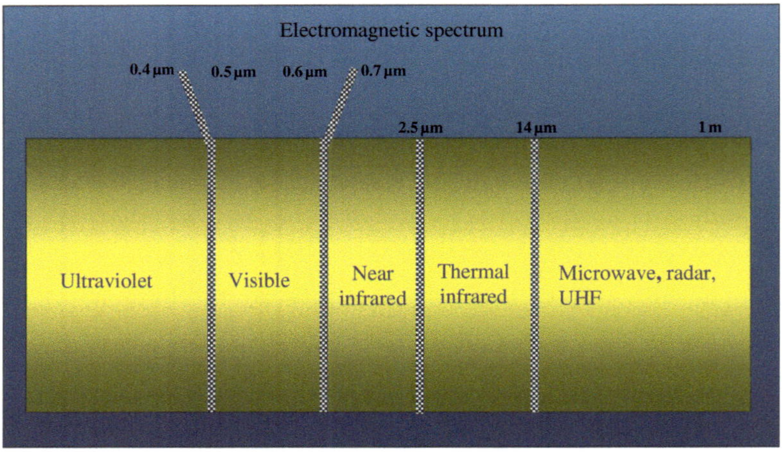

Figure 1.1 A portion of the electromagnetic spectrum.

1.2 FILM

The main advantage of focusing on the visible portion of the spectrum is that film emulsions, color and black-and-white, are available there. Why is this important?

1. There are many commercial aerial photography firms that use film.
2. Time-lapse aerial photo coverage is available for many areas (dating back to the 1930s in some areas).
3. Films are inexpensive ($15–20 per exposure depending on number of exposures).
4. Films have high geometric fidelity (few distortions).
5. Colors and gray tones are familiar to us. After all, we have used the visible portion of the electromagnetic spectrum since birth.
6. Films have high spatial resolution.

Although aerial photography firms throughout the world still use film and camera systems, many modern firms also offer high-resolution digital aerial imaging. Although the digital acquisition systems sometimes use scanners, the geometric fidelity is excellent on modern systems. Digital products provided for interpretation, however, most often involve scanning the digital data onto film. Hence the advantages of digital imaging in the visible portion of the spectrum are essentially the same as those mentioned above.

1.3 TARGET INTERACTIONS

So what happens when sunlight strikes a target? Three things happen as follows:

1. *Sunlight is reflected back from the target surface.* Once reflected, the energy can be sensed by a camera, your eye, or a scanner of some kind. The amount of energy reflected from a target is determined by the *reflectance* of the target. Some colors may reflect more energy than others. If most of the reflected energy is in the red portion of the spectrum, for example, the object will have a reddish hue. The color of a target depends on its *spectral reflectance* (reflectance in different portions of the spectrum).
2. *Sunlight is absorbed by the target.* Absorption occurs when a photon (packet) of energy excites atoms at the surface of the target forcing electrons to jump away from the nucleus into more distant orbits, thus absorbing the energy of the photon. This new atomic arrangement keeps the energy within the surface atoms, and prevents reflection. It also causes the atoms to vibrate more rapidly, increasing the thermometric temperature of the surface. Usually the new electron configuration is relatively unstable, and the electron will eventually drop back into its original orbit. When this happens, the energy originally absorbed is emitted by the surface atoms at longer wavelengths, frequently in the thermal infrared region. The amount of sunlight absorbed is determined by the target's *absorptance* and the amount of energy absorbed by different colors its *spectral absorptance.* Iron minerals, for example, absorb strongly in the blue and green portion of the visible spectrum, thus little blue and green energy is available for reflection. This results in the reddish appearance of iron minerals such as iron oxides.
3. *Sunlight is transmitted by the target.* Transmitted energy is propagated through the target where it may interact with another object below. The energy transmitted is determined by a target's *spectral transmittance.*

There may be some confusion about the terms *reflectance*, *absorptance*, and *transmittance* used above relative to the terms *reflection*, *reflectivity* and similar terms for *absorption*, *absorptivity* and *transmission* and *transmissivity*. I'll try to clear this up using the reflection process and expanding to the others intuitively.

Reflection is a *process* in which energy is reflected from a surface, the amount of energy reflected depends on the wavelength of energy, the roughness of the surface, the angle of incidence of the energy at the surface, and the electrical properties of the surface (reflectivity), among others.
Reflectivity is an intrinsic property of a *material*. It depends on the electrical properties of the material. It is a property that one can find in the *Handbook of Physics and Chemistry*. Gold, for example, has a constant reflectivity regardless of the character of its surface.
Reflectance is an intrinsic property of a *surface*. It depends on the reflectivity of the material composing the surface, and the other surface properties mentioned above. The reflectance of a smooth, pure gold surface, for example, may vary from the reflectance of a corroded or abraded gold surface.

Look down at the earth for a moment. Can you see into it? I can't. So I think we can safely say that for most earth materials we deal with at aerial photography scales the transmittance is zero. Then

$$\rho(\lambda) = 1 - \alpha(\lambda)$$

In this equation, $\rho(\lambda)$ is the spectral reflectance and $\alpha(\lambda)$ is the spectral absorptance, at wavelength λ (*color*). This is a trivial equation that says that if the electromagnetic energy is not reflected at a surface, it is absorbed.

For some vegetation, shallow water receiving differential sediment input or having different carrying capacities or turbidity, and snow and ice fields, one can occasionally sense a short distance below the actual surface. We will discuss these instances in more depth in Part III (Chapters 9 through 12).

1.4 MORE ABOUT RESOLUTION THAN YOU PROBABLY CARE TO KNOW

Resolution is all about measurement accuracy. We will need this mostly in Part III (Chapters 9 through 12) when we deal with imaging beyond the visible portion of the spectrum. There are really three kinds of resolutions:

1. *Spatial resolution*, which deals with being able to distinguish two objects located close together in space as two distinct objects rather than one.

2. *Signal resolution*, which deals with being able to distinguish two image amplitudes (brightnesses) as two distinct brightnesses rather than one.
3. *Spectral resolution*, which deals with being able to distinguish two colors (hues), as two distinct colors rather than one.

1.5 SEEING THE WORLD IN 3-D

There are additional advantages to using film. The use of film implies employing a camera system, where discrete photographs are taken sequentially along a flight line. When two such photographs are viewed simultaneously with the aid of a stereoscope, the separate images merge into a single three-dimensional (3-D) image. A photo analyst who sees the world in 3-D will outperform a 2-D analyst every time. This e-book will concentrate on the use of 3-D interpretation in the visible region.

In regions beyond the visible strange things happen:

1. Rock and soil may behave differently when irradiated there.
2. Film emulsions may not be available there.
3. Atmospheric effects may differ there.
4. Brightness may imply different processes there than in the visible.
5. Surfaces that appear rough to our eye may appear smooth there.

So, you see, there are good reasons to begin our exposure to aerial interpretation by studying visible patterns on photographs. Later, near the end of the e-book, we will take our knowledge of normal patterns and attempt to extend their validity to portions of the spectrum beyond the visible.

1.6 COLORS AND PATTERNS AND SHAPES—OH MY

We have been learning colors and patterns and shapes virtually since we took our first breath. Just about everything we have learned, from the look and smell of a rose to the printed symbols that you are reading at this moment, has presented us with a pattern recognition challenge. Throughout our life we have, through a long, laborious and sometimes painful process of experimentation and/or education, stored in our memory combinations of colors and patterns and shapes. We learned about our surroundings by associating these colors and patterns and shapes with our knowledge about—*things*. Once learned, the colors and patterns and shapes, together with their associations with things, can be recalled quickly whenever they are encountered again. This association of colors and patterns and shapes with things is the essence of interpretation.

Look back now at the aerial photograph in Figure I.1 (page 2). I made the remark that what you *think* you see in the photograph depends on your experience with aerial photographs. That is true of any pattern. We *see* what our brain *tells* us we see; what we have been trained to see. You recognize the symbols on this page readily, I hope, and perhaps you have reflected back on other things you have learned to gain some insight into the human visual and neural processes. It is to be hoped that your understanding has gone way beyond what can be gained by simply recognizing these symbols, and that the meaning or interpretation of the symbols (letters and words) has suggested new ideas (that is, you have *read between the lines*).

What happens now when we are confronted by new patterns, patterns different from those we have learned? Look for a moment at the patterns in Figure 1.2.

There is probably little your brain can do to help you interpret these patterns unless you have been trained to recognize Hebrew or Arabic characters. If you wish to learn these patterns for use later, you must follow the simple two-step process presented below:

1. Store the patterns. The more you store, the better you will become at understanding Hebrew and Arabic.
2. Associate the patterns with *things* in your knowledge base. You obviously need to do more than merely recognize a pattern as one you've seen before; you also need to understand what the pattern means.

גאגדהכלןףפץצקרשתווק

فقظضمكنيئق٣ؤظسشزكل

Figure 1.2 Randomly selected characters from Hebrew (top) and Arabic (bottom). No attempt has been made to create words. I hope I haven't inadvertently created any offensive words.

This is the same two-step process you will need to become good at interpreting aerial photographs. This e-book will provide you with landform patterns commonly found on aerial photographs and I will share with you the things (geologic processes) that naturally form the patterns. The more patterns you store (the more time you invest), the better you will become at interpretation.

Obviously, I can't cover all possible geologic or landform realizations in this e-book. So it will be up to you to continue the two-step process of learning colors and patterns and shapes, and associating them with the things that are important to your interests and applications. The way to do that is to *use* aerial photographs. They are cheap and valuable. You can't possibly get more bang for your buck. Aerial photographs should be the first thing you think about during project planning, and the last thing you put away as you write your final report. Remember, there is no excuse for missing something dangerous that you could have seen on an aerial photograph. Doing so will cause you no end of grief in court even if it doesn't cost human lives.

1.7 READING AERIAL PHOTOGRAPHS

All right, enough talk. It's time to take a closer look at an aerial photograph. Before we become overly involved, though, let's relax and spend some time just looking so that we can develop a flexible strategy for our ultimate interpretation. Look back at Figure I.1 (page 2). It may be a good idea to print that page, or use the bookmark provided, so that you can refer to it while we continue our discussion.

A general strategy that I use whenever I visit a site on the ground also helps me interpret aerial photographs. It goes like this:

> *Sit down (preferably on a hill where you can see most of your site). Don't immediately begin writing. Quietly observe for 30–40 min. As you scan the area you will find that your first impression of being overwhelmed by the size and complexity of the site and its environs subsides. As it does, you begin noticing things that your mind wouldn't allow you to see when you first arrived. The longer you observe (you can repeat the exercise at different times of day as well), the more subtleties of the site become apparent.*

The same strategy works well for interpreting aerial photographs. Avoid immediately becoming preoccupied with small details of your image. Stand back, so to speak, and look at Figure I.1 for several minutes.

As you look at the photograph, notice that it can be conveniently divided into four regions as follows:

1. the light, reddish-brown area at the top of the photograph;
2. the dark green striped area at the far upper right corner;
3. the dark brown, mottled area at the lower right corner; and
4. finally, the light brown area with little texture occupying the rest of the photograph.

Once convenient segments having similar characteristics have been delineated, we can begin making more detailed observations within each region. Observations usually fall into the following seven categories:

1. *Color, or tone* on black-and-white photographs, refers to brightness, spectral combination, or amplitude information. It is one of only two unique kinds of information that exist on a photograph. It is a qualitative measure of the total amount of sunlight reflected by the ground surface in the different portions of the visible spectrum of Figure I.1 (page 2). The bright surfaces seen on the photograph are reflecting more sunlight, whereas dark surfaces reflect less (absorb more). Later we will see that color or tone carries the most dependable information on the *type* of rock or soil composing the surface. Notice that the darkest tones on the photograph occur in the bottoms of the two streams at the center of the photograph. Does this mean that water absorbs most of the sunlight? Well, not exactly. Water certainly absorbs some sunlight, or your swimming pool wouldn't heat up in the summer. It transmits sunlight also, something I neglected to mention earlier when talking about the reflection of the ground. Rock and soil do not effectively transmit sunlight; if they did you could see through them. No, water reflects a great deal of sunlight. This apparent discrepancy is caused by the roughness of the surface. Rock, soil, and vegetation have surfaces that are rough enough to reflect sunlight in all directions (called Lambertian reflection). Water, however, often has a smooth surface so sunlight reflects from it as it would from a mirror (called specular reflection). There is plenty of sunlight reflecting from the water in the bottom of those streams, you just won't see it unless the camera is at just the right angle. The darkest tones on most photographs are caused by water, shadows, and vegetation, the lightest by rock and soil.

2. *Texture* refers to the changes in color or tone as one scans across the photograph. Color (or tone) and texture are the only two unique kinds of information available on a photograph. Notice that the texture in the lower right corner of the photograph differs from that in the center of the photograph. Colors that change rapidly can be called high-frequency textures, whereas those changing slowly are low-frequency textures. *Texture relates to rock hardness and geologic structure*; it will be our best indicator of potential natural hazards.
3. *Shapes* are caused by boundaries around relatively continuous colors or textures. On aerial photographs shapes are often useful to delineate cultural activity (the dark green agricultural fields in the upper right corner of the photograph are examples), geologic features (the reddish cinder cone to the left of the lowest agricultural field is an example), and erosion.
4. *Pattern* describes the ways that colors or textures combine to form ordered sequences. I have previously used the word "pattern" to refer to a generic type of problem (pattern recognition), from here out I shall restrict the word to describe the patterns created by drainage systems.
5. *Size* is relative; it depends on the scale of the photograph. You may, for example, alter your interpretation of a dark-colored circular feature from a farmer's centripetally irrigated field to a cinder cone, depending on its relative size.
6. *Shadows* are wonderful for enhancing texture. Judicious choice of time of day or season of year for acquiring aerial photographs makes it possible to preferentially enhance textures having different orientations. The fault traversing the left edge of the cinder cone is visible mostly because of the shadows produced by accelerated gully erosion on the upthrown (left) side of the fault. Be aware that most commercial aerial photography firms make their money by producing aerial photographs for topographic mapping projects. In producing topographic maps, shadows are anathema, so, unless otherwise instructed, commercial aerial photography firms conduct flight operations near midday, thus washing out important texture information. If you want maximum texture in your aerial photographs—and trust me you do—you will have to instruct the contractor to fly in the morning or afternoon.
7. *Location* of an object or feature can lead to important interpretations. For example, the dark-colored swampy area between the two

streams mentioned earlier on the left of Figure I.1 is unusual. Why would one find a swampy area on a sloping hillside? A butterfly? I think so.

Combinations of these seven observational elements, sometimes called "*interpretation keys*," are used to make interpretations. You will seldom find it necessary to actually list them, but you will use them all subconsciously as you become a better photo analyst. Remember, there is a big difference between observations and interpretations. The following are observations on Figure I.1:

1. There is a lack of a well-defined drainage pattern and few water courses.
2. There are dark green colors in the upper right corner of the photograph.
3. There is a truncated reddish circular shape near and to the left of the green-colored area.

These are all valid observations and one can make them all day without fear of criticism. Unfortunately, it is unlikely that you will find anyone willing to pay you to make them. What you will be paid for is sticking your neck out and making an interpretation as to what these observations suggest in terms of hazards. For example, the light color and low stream density are indicative of young, sandy sediment with high permeability. The truncated reddish circular shape is a volcanic cinder cone. It is truncated at its western edge (left) by a normal (dip-slip) fault. The joint occurrence of volcanism and faulting is not uncommon, but the truncation suggests that the most recent activity on the fault postdates (is more recent than) the cinder cone. Furthermore, the fault rupture of young alluvium indicates a dangerous active fault. Any engineering designs in the area should avoid active fault scarps (there are others visible on the photograph—can you see them?) and be consistent with good engineering practice anticipating strong ground motion.

1.8 SUMMARY

So there you have it. When you sit down to interpret an aerial photograph, begin by relaxing and simply appreciating the photograph. Let your gaze sweep the entire photo area without serious purpose. Next divide the photo area into large regions having similar characteristics (divide and conquer). Jot down some observations for each region. Use the interpretation keys as a guide, but don't feel constrained by them. Finally, make an interpretation based on your knowledge, experience, training, and observations.

If your eyes get tired, or you find yourself scribbling down notes, put the photographs down for awhile and move on to something else. Come back to the photos later. You will find that you will see something new and interesting every time you view the photographs. Looking at pictures is fun. Enjoy the process. When you stop having fun, put them down until later.

It's really all about efficiency and due diligence. Why would anyone want to stumble around on the ground for hours, perhaps looking in the wrong places, and missing important features when *most* of the features of interest can be calmly delineated on an aerial photograph in an air-conditioned office? Do you like trudging up and down hills and wading through cactus? Make notes on the photographs indicating places of particular interest to visit on the ground, or places that are confusing or lack sufficient detail on the photographs. Doing so provides important focus for your field study.

Take the photographs into the field with you. Relax in a high, shaded locale for 30–40 min and orient yourself with respect to the aerial photographs and the site. Annotate your photographs using clear acetate overlays to protect the photo surface. This is an iterative endeavor best accomplished by performing repetitive cycles of field reconnaissance/mapping and photo analysis (the two-step learning process described on page 8, above). Unless you want different people learning different things, then arguing about them later, the photo analysis and field reconnaissance/mapping should be done by the same individual.

By all means, though, share your interpretations with colleagues. A lively discussion of interpretations is stimulating. It is also acceptable to have more than one possible interpretation for ground features. These can usually be resolved on the ground at the site. If one interpretation is wrong, shift to an alternate. I told my students that it is acceptable to be wrong—what you always want to avoid, though, is being surprised.

CHAPTER 2

More About Photographs Than You Ever Cared to Know

My mission is to help you learn to recognize hazardous ground conditions using aerial photographs. Before we begin there are some things that are useful to know about the photographs themselves. Not many—but some. What we will discuss now will enable us to make accurate measurements on aerial photographs and to understand three-dimensional viewing. Stay with me on this, we will return to interpretation soon. Or, if you wish, skip this chapter and come back to it if you need to later.

2.1 PHOTOGRAPHS

Figure I.1 shows a typical layout of an aerial photograph. Typical negatives are 254 mm (10 in.) wide and have an image size of 228.6 mm by 228.6 mm (9 in. by 9 in.). At the top of most aerial photographs you will find some information about the photo mission. The format will differ depending on the camera and needs of the contracting organization. From left to right on Figure I.1 is the date the photograph was taken, the scale (1:24,000), the contracting agency (US Bureau of Land Management), project identification, and a photograph identification number. Usually, although not on Figure I.1, you will find a circular level bubble.

On the frame of the film canister, or on the camera body, are marks (usually black) located in the corners and midway between the corners of the picture. These marks are called *fiducial marks*. The purpose of fiducial marks is to allow you to find the precise geometric center of the photograph. This center is called the principal point. Consider Figure 2.1 for a moment.

The spot on the ground directly below the center of the camera lens will make an image on the film located at the focal plane of the camera. Assume there is a black rock at that point. We will see the black rock on the photograph. If the camera is looking vertically down at the ground as

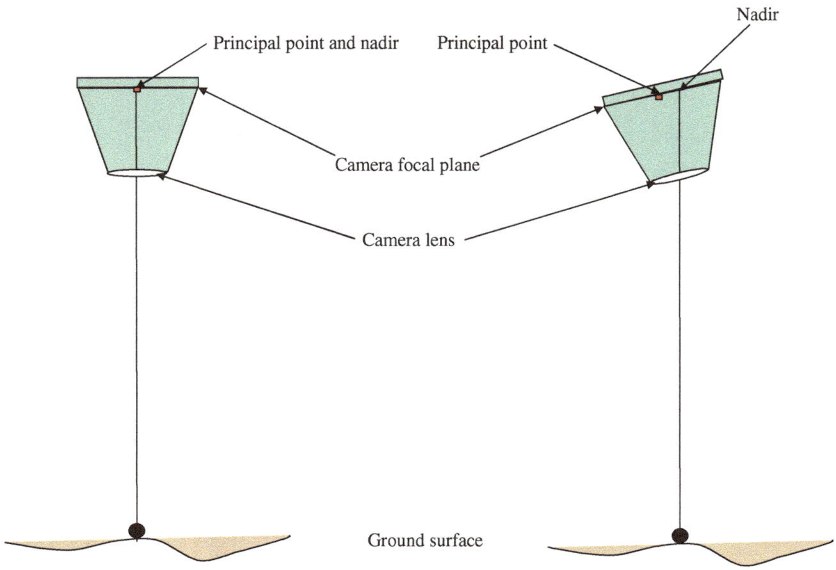

Figure 2.1 This figure shows the definition of nadir vis a vis principal point.

shown on the left of Figure 2.1, the black rock and the principal point will coincide. The point under the rock is called the nadir point. The nadir point always coincides with the principal point if the photograph is vertical. When we make measurements on aerial photographs, we assume that the photos are vertical so that when we find the principal point (a simple process of connecting fiducial marks with straight lines and noting their intersection) we are confident that it coincides with the nadir point.

Occasionally, as on the right of Figure 2.1, the camera is not vertical, but rather oblique to the ground. When this happens the nadir point and the principal point no longer coincide. If you can locate the nadir point in a situation like this you can theoretically make measurements on the photograph—but I wouldn't advise it.

Well fine, I know you're thinking, if we don't have some helpful soul putting black rocks below the camera lens, how can we tell if the photo is vertical? There are a couple of ways. If we are lucky, the camera is so oblique to the ground that we can see the horizon on the photograph. If you can see the horizon, the photograph is called a *high oblique* photograph. Don't make measurements on these. If the horizon doesn't appear on an oblique photograph, it is called a *low oblique* photograph.

Don't make measurements on these either. If in doubt, take a look at the level bubble. As long as the bubble is within the central area, your measurements equating the nadir and principal points should be acceptable. A bubble that is squashed against the side threatening to escape is telling you not to make measurements on the photograph.

In practice, commercial aerial photography companies perform excellently; they do not take oblique photographs unless specifically asked. Why would anyone bother to ask? Well, oblique photographs provide a more natural view for humans than do vertical photographs, especially high oblique ones. I *have* witnessed some people walking around with their head staring vertically down at the ground, but they always have scars on their forehead, so it is generally considered a practice to avoid. High oblique photographs make great panorama shots helpful in gaining an overall understanding and appreciation for the area of interest, or to help illustrate a point in a report. I routinely snap off a few high oblique photographs for this purpose. At the end of a flight line, I will keep the camera running and ask the pilot to bank sharply to bring the camera up parallel with the horizon. As the aircraft reaches its steepest bank, the high oblique photographs are taken. On the way up and down low oblique photographs are collected. My stomach never forgives me for such a practice, but the pilots always seem to enjoy the break in tedium.

2.2 HOW BIG IS BIG?

The first measurement people invariably want to make on an aerial photograph is the size of something. What is the diameter, for example, of that cinder cone on Figure I.1? If we photograph the cinder cone from a height of 100 m, I guarantee it will appear much larger than it will if we use the same camera lens but increase the height to 6000 m. So the size of something on an aerial photograph depends on its actual size on the ground, the camera's focal length, and the height of the camera above the object. It depends on the photograph's *scale*. The scale of a photograph is defined as

$$\text{Scale} = \frac{\text{fl(m)}}{H\text{(m)}}$$

where fl is the focal length of the camera (length from center of lens to focal plane, refer to Figure 2.1), and H is the height of the aircraft above the ground, all measured in the same units (meters here). Stop here and notice a couple of things that may seem trivial but are the sources of much confusion. First, the scale is a dimensionless ratio. It is represented most often in two ways:

1. 1:24,000
2. 1/24,000

You can obtain the scale in this form by multiplying the numerator and denominator of the scale equation by the reciprocal of the focal length (1/fl). The scale is read "*one to twenty four thousand*," and implies that 1 mm measured on the photograph corresponds to 24,000 mm on the ground. For convenience, some map makers will convert the scale to read "*1 in. (on the photograph) equals 79 ft (on the ground).*" This may indeed be convenient, but it is not standard and will not be used further. If you have a photograph from which the scale information has been removed, you can still compute a scale if you have a map of the area. Then

$$\text{Scale} = \frac{\text{DP(m)}}{\text{DG(m)}}$$

where DP is the distance between two objects on the photograph, and DG is the distance between the same two objects on the ground, measured in the same units. If using this relationship try to choose points on the photograph on opposite sides of the nadir, because on wide angle lenses distortion occurs at the edges and corners of the photograph.

2.3 PHOTOGRAPHS VERSUS MAPS

Most of us are familiar with using maps. Measuring distances on maps is easy. We compare the distance between the tip of our thumb and the first joint of our thumb with the printed map scale bar, and merrily go about making bad measurements. Those measurements are bad on maps, but they're even worse on aerial photographs. On an accurate planimetric map, all features are depicted at their correct spatial positions and the observer has a true vertical view of every detail shown.

This standard cannot be met by aerial photographs due to various sources of distortion or image displacement. Objects observed on aerial photographs may fail to register in their proper planimetric positions due to a number of factors, some of which are given below.

1. Optical or photographic deficiencies—optical distortion is due to an inferior camera lens and the recorded images display distortions both away and toward the nadir. Faulty shutters, film shrinkage, or failure of the camera vacuum can also cause image distortion. With the precision mapping cameras used today, this type of problem seldom occurs.
2. The exact angle and direction of camera pitch and roll are seldom known to the interpreter. Tilt amounting to 2–3° or less can usually be ignored without serious consequences.
3. The most significant source of image displacement on aerial photographs is topographic relief. Some parts of the ground are simply closer to the camera than others. Consider Figure 2.2. On an aerial photograph, the building appears to lean radially outward from the nadir. The problem is that if we want to know the distance between two buildings, do we measure the distance between the bases or tops of the buildings? Remember the bases may also be at different elevations.

Actually the occurrence of image displacement (*parallax*) due to topographic relief is more a help than a hindrance, because it enables us to see the world in three dimensions. The higher the object, the more it will be displaced on the photograph. From most aerial photo

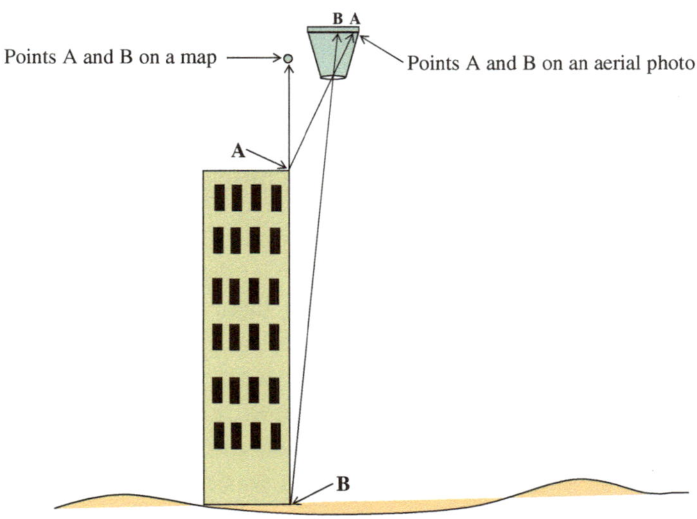

Figure 2.2 This figure shows an example of image displacement due to elevation.

companies, it is possible to obtain products called *orthophotographs*, which have been corrected for parallax and have the same spatial fidelity as a map. Orthophotographs are fine for architects, realtors, planners, or politicians, who wish to make measurements, but they are not stereoscopic images. If you are interested in recognizing natural hazards, stick with the original aerial photographs and interpret them using stereoscopic vision.

2.4 WHAT IS STEREOSCOPIC VISION?

Now it's time for the most enjoyable part of interpreting aerial photographs. The reason we can see in three dimensions is that we have two eyes and our eyes are offset from one another. If there are any one-eyed people reading this, don't despair, you can still use aerial photographs to recognize dangers and hazards on the ground, but understand that you will always be at a disadvantage with respect to an analyst who can see in stereo.

Recall that one of the most important benefits of using aerial photographs in the first place is that individual photographs are taken along a flight line. Aerial photography companies will take photographs close enough together so that, at a minimum, 60% of adjacent photographs image the same ground area. If you now look at a pair of adjacent photographs so that one eye looks at one while the other eye looks at its stereo pair, the image displacement described earlier will make it appear to be a single 3-D photograph. The trick is being able to train your two eyes on individual photographs. After some practice, you will be able to do this without any aid. Until then, however, you can purchase an instrument called a *stereoscope*. Using mirrors or oriented lenses, stereoscopes focus your eyes onto individual photographs.

Before proceeding, print out the stereo pair of aerial photographs on the next page, turn the page on its side (landscape format) and use a stereoscope to view the images until you have convinced yourself that you can see in stereo. My eyes will automatically fuse the stereo pair into a single 3-D image, but you may find it easier if you cut and separate the photos so that you can adjust their separation. This is important for your ultimate success as a photo analyst. If you have trouble at first, try again. One thing that always helped my students was to be sure that you relax your eyes as if you were looking down at the earth from an airplane. Sometimes your eyes insist on focusing on the near field instead of on the far field. Another useful trick is to place your left index finger on the cinder cone on the left photograph and your right index finger on the cinder cone on the right photograph. Now look into the stereoscope and adjust the photo separation until your index fingers merge. Eureka! My most gratifying moments as a teacher occurred the moment one of my students was first able to see an image in 3-D.

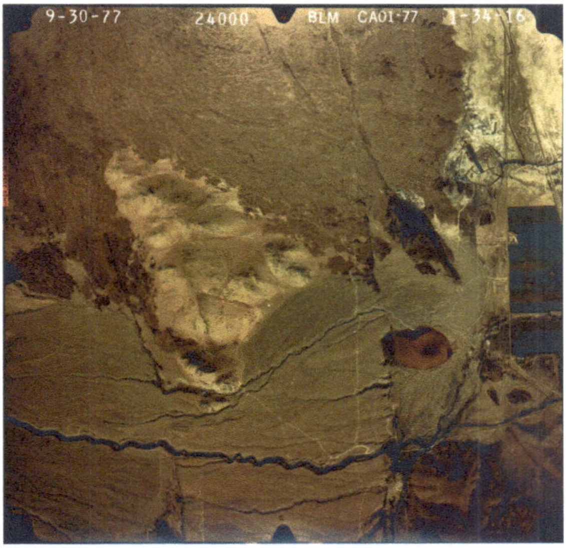

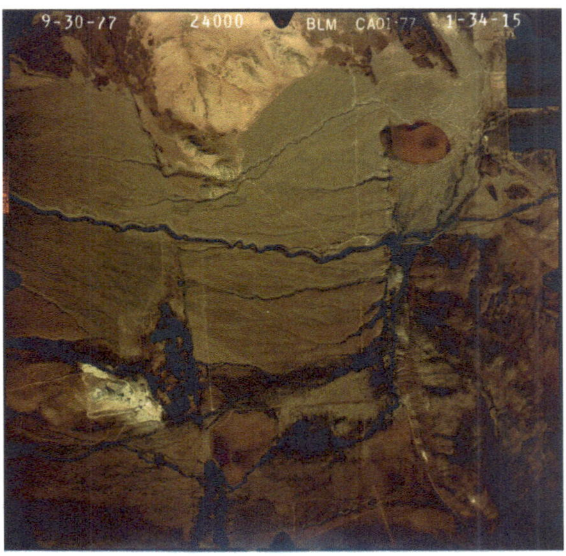

Here's another word of advice. To see proper stereo, your stereo pair of photographs must be arranged parallel to the flight line, as shown in the following figures. If your alignment looks like the combination below, you are doing it wrong.

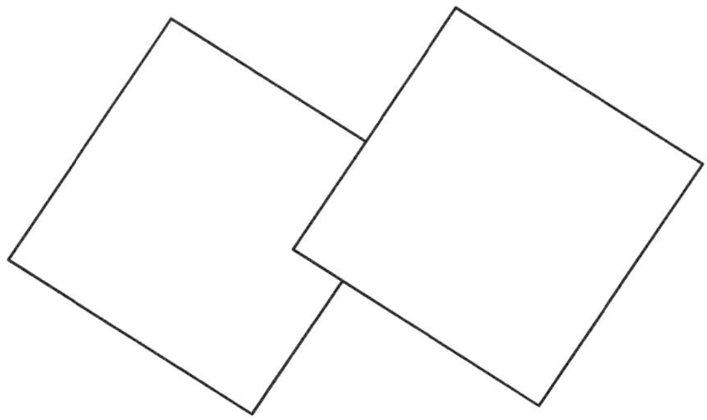

In stereo you can now appreciate the relief demonstrated by the small cinder cone (north of Red Mountain in Owens Valley, California) and the size of the offset on the fault scarp that truncates it. A *fault* is simply a fracture in the earth across which movement has occurred. A *fault scarp* is a fault that ruptures the surface of the earth. Not all faults do. You can also see a fault scarp upstream from the cinder cone. This scarp does not influence the youngest alluvial material as does the downstream one, so rupture on this scarp is older. Still, it displays a significant effect on older surfaces and on the streams that cross it. Note, for example, how streams flowing from the west divert at the older fault scarp, some actually flow along the scarp for a distance before resuming their eastward trend. The fault scarp truncating this cinder cone ruptured in 1872 causing the magnitude 7.6–8.0 Lone Pine, California earthquake. That earthquake was one of the largest to affect the conterminous United States.

Don't proceed further until you have acquired stereo vision of the earlier photographs.

2.5 HOW MANY AERIAL PHOTOGRAPHS DO I NEED?

Obviously, before we can look for dangers in our area of interest, we need to purchase some aerial photographs or, if none is available, have some taken. Since this costs money, it is a good idea to have some idea of how many photographs will be needed to provide full stereographic coverage of our area. Then we can budget for the coverage in our project. Remember, aerial photographs will save you an enormous amount of time in the field, so the cost is rather easily justified in man-hour savings alone. Also be aware that photo mosaics available from aerial photo companies seldom possess full 60% stereo coverage.

The calculation is really quite simple, but for some reason whenever my students tried it on an exam, several of them fumbled it. There are three parts to the calculation, and some information is needed first. Necessary information includes:

1. Photo scale—The photo scale is an important piece of information. Large scale photographs provide more ground detail, but it will take more of them to cover a given area. Small scale photographs cover a larger area, hence it will take fewer to cover the site, but you lose some ground detail. I have found a scale of 1:12,000 to be optimum for most cases, but to keep the math simple let us assume we want a scale of 1:10,000. Remember that photo scale is a ratio, so that 1:12,000 is the same as 1/12,000, a smaller number (scale) than 1/10,000.
2. Photograph size—Most mapping cameras provide an image size of 9 in. × 9 in. (228.6 mm × 228.6 mm).
3. Size of area of interest (let us assume 50 km × 25 km for our example).
4. Amount of sidelap (percentage of overlap between flight lines, say 30%).

With this information we can estimate photo cost.

2.5.1 Part 1: Ground Covered by One Photograph

The first part of the calculation is to use the photo scale and photo size to compute how much ground is covered on a single photograph. The photograph is 228.6 mm by 228.6 mm. The photo scale of 1:10,000 implies that 1 mm on the photograph corresponds to 10,000 mm, or 10 m on the ground. So, one photograph covers a ground area of 2286 m by 2286 m, or 2.286 km by 2.286 km.

2.5.2 Part 2: Ground Advance Per Photograph

Since each photograph covers 2.286 km by 2.286 km, if we were to plan our photo project so as to provide no overlap, each new photograph would cover a new 2.286 km by 2.286 km patch of ground.

That is, the principal points of adjacent photographs would be separated by 2.286 km. So we might say that each photograph *advances* us 2.286 km along the flight line. For stereoscopic coverage, though, we want the photos to **overlap** by 60%. This means that our advance along the flight line drops from 100% to 40%. Sounds trivial, I know, but this is the biggest source of errors. The ground advance along the flight line is then 2.286 km $\times$ 0.4 = 0.9144 km.

This is also a good time to compute the advance from one flight line to another. We obtain all of our stereo coverage using overlap, so we don't need sidelap for that purpose. Sidelap is used only to make sure there are not accidental gaps in coverage. It is not unusual for the pilot to wander a bit off line because of winds or lack of good landmarks to keep the plane on line. So the amount of sidelap is arbitrary. Our choice of 30% is a bit conservative, 20% is more common, but let's go with it anyway. Side advance from one flight line to another is 2.286 km $\times$ 0.7 = 1.6 km.

2.5.3 Number of Photographs

The number of photographs per flight line depends on whether we fly lines along the long axis of the area of interest (50 km) or the short one (25 km). Practicality will dictate this. To minimize the amount of flight time off the target (i.e., the number of turns and line-up time), lines will be flown along the long axis, so

$$\frac{50\ \text{km/line}}{0.9144\ \text{km/photo}} = 54.68\ \text{photos per line}$$

Now, we need an integer number of photographs so always **round up**. In addition, I will add one or two just for insurance, so we need 57 photographs for each flight line. The area of interest is 25 km wide, so

$$\frac{25\ \text{km}}{1.6\ \text{km/line}} = 15.62\ \text{lines}$$

Again, we want an integer number of lines, so round up to 16 lines, but ***do not*** add lines for insurance. Each additional line adds 57 photographs to the total. Now is a good time to add a few of those oblique photographs we discussed earlier. If you want some, add them (say, 10).

2.5.4 Total Photographs and Cost

We have 57 photographs per line, 16 lines, and 10 obliques, for a total of 922 photographs. At a cost of approximately $20 each for color photographs, the total cost is approximately $18,440. To reduce the cost, we can trim the sidelap, opt for black-and-white (panchromatic) photographs, or decrease the scale. ***Do not alter the overlap***.

2.6 CHOICE OF COLOR VIS A VIS PANCHROMATIC

If the budget permits, I will always choose color photographs over panchromatic. Color photography is optimally done during the middle of the day when color information is at a maximum. Most natural hazards, however, are revealed on aerial photographs through textural information. Textural information is enhanced by shadows, which occur early or late in the day. Hence, we are faced with a tradeoff—collect color photographs at nonoptimal times, or collect panchromatic photographs and sacrifice the color information. My choice would be the former—collect color photographs in the morning or afternoon to enhance texture while still benefiting from the superior information provided by color over panchromatic photographs.

Look back to the stereo photographs of the 1872 fault scarp on page 21. You see few shadows even in the gullies eroded by the streams, indicating the mid-day flight time. If the photographs had been taken later in the afternoon, you would see shadows on the fault scarp and enhancement of the extensive graben and fissure development on the down-thrown (east) side of the scarp. You would also see more textures related to the fault farther west.

CHAPTER 3

Textural Information

3.1 STREAM PATTERN

Streams and cultural activity are the primary sources of texture on aerial photographs. Some of the dangerous conditions of concern to us may be caused by cultural activity. This activity can usually be detected directly by the textures they produce (roads, trails, excavations, mining activity, etc.), or frequently by the subtle effects these activities have on streams. We will first look at natural stream erosion so that we can address departures that may reveal natural hazards. There are three important aspects to stream erosion:

1. *Stream pattern* indicates geologic structure and changes that may reveal dangerous ground conditions.
2. *Stream density* indicates grain size and hardness of surficial materials.
3. *Stream bed and gully characteristics* indicate hardness of surficial materials.

3.1.1 Dendritic Drainage Pattern

Consider this simple experiment. A homogeneous layer of fine silt is placed in a shallow box. At the top of the box is affixed a trough. This trough has a long narrow slit having the same length as the trough itself. Now the box is placed at a small angle, the trough filled with water, and we observe the flow of water over the silt. The initial flow spreads out to form a sheet of water.

After water flows for some time, series of shallow, parallel grooves (rills) form in the silt, essentially down the regional slope. These rills are erosional and mark the beginning of a drainage system. By numerous mechanisms, the rills deepen and unite. During rill erosion, there is a tendency for angular bends in rill systems to develop into rounded meanders due to lateral thrust of current on the outside of curves. As long as the surface remains gentle and the water supply is maintained, this process of rill merging, deepening, and curving continues.

The resulting pattern looks like a tree with its trunk at the downstream end. Because of the treelike pattern, it is called a *dendritic pattern.*

A dendritic pattern appears to be nature's pattern of choice; perhaps it represents a minimum energy pattern. It is the most common drainage pattern found on a regional scale and is characteristic of essentially flat-lying to gently dipping, homogeneous materials. It can be subdivided according to degree of tributary density into *coarse*, *medium*, and *fine* densities. A coarse dendritic pattern is found on coarse grained or hard surficial materials. A fine dendritic pattern is found on fine grained or soft surficial materials.

Bends in a dendritic pattern are natural and are probably due to water quantity, stream gradient, and minor asperities or perturbations. Alignment of bends can easily be fortuitous and are not necessarily indicative of natural hazards. A dendritic pattern is, however, nature's pattern of choice; it is the norm, the indicator for natural conditions. *If you don't have a dendritic pattern, it's indicative of a departure from the norm, a butterfly, and there must be a good reason for it (usually structural control). Search diligently for that reason, and always pay close attention to it.*

Figure 3.1 shows a cartoon depicting a dendritic drainage system in which the stream flows from upper left to lower right.

There are a couple of things to note before we see how things act to alter a dendritic pattern. First, a dendritic pattern is *symmetric*, so that

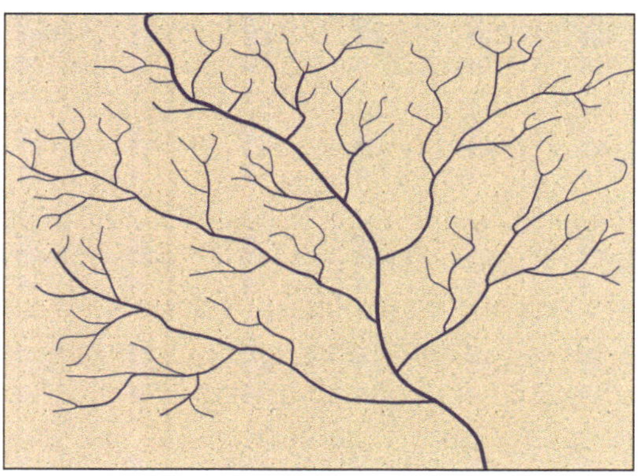

Figure 3.1 The figure shows a typical dendritic drainage pattern. Adapted from von Bandat, 1962[1] and Strandberg, 1967[2].

stream lengths on both sides of the main stream are virtually equal. Second, as streams merge with larger streams downstream, they do so at acute angles. Third, the drainage area is well drained, which means that streams continue progressively down-slope to larger streams. There are no sinks or areas where streams disappear. Remember that *a dendritic pattern tells us nothing about the rock or soil type upon which it develops*. Shale, sandstone, or granite all may display a dendritic pattern. Clues as to rock or soil type are revealed by drainage density and gully characteristics, which will be discussed later.

3.1.2 Parallel Drainage Pattern

If the slope of our tilted box becomes steeper, the dendritic pattern will elongate as shown in Figure 3.2. Note that the symmetry, full drainage, and angle at which streams join higher order streams are maintained as the dendritic pattern is elongated by the steeper slope. We are still dealing with a relatively homogeneous surface so these characteristics of a dendritic pattern are maintained. As the slope becomes even steeper the elongation of the dendritic pattern continues until the increased energy of the water overwhelms minor irregularities, and flow in nearly parallel straight lines down the slope. Such a *parallel pattern* is shown in Figure 3.3. Parallel drainage patterns are indicative of regional sloping terrain conditions, or a system of parallel faults or joints. Parallel patterns also occur on alluvial aprons.

Figure 3.4 shows a panchromatic vertical aerial photograph of an area in west Texas. The photo displays predominantly dendritic drainage patterns. Streams flow from a high mesa, beyond the right edge of the photograph, to a large stream at the upper left. The dendritic pattern is indicative of relatively homogeneous surficial deposits gently

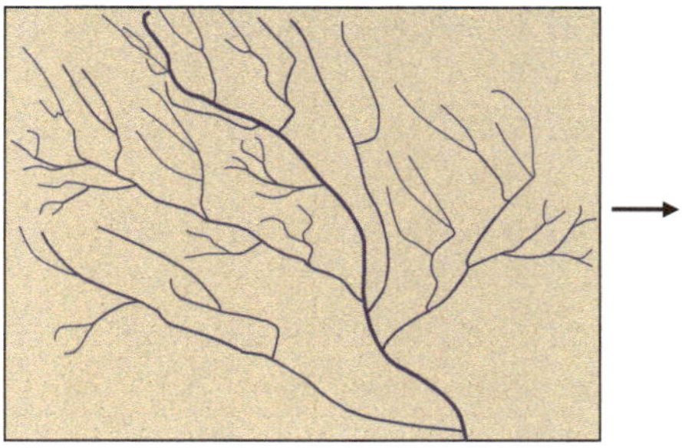

Figure 3.2 Elongated dendritic pattern.

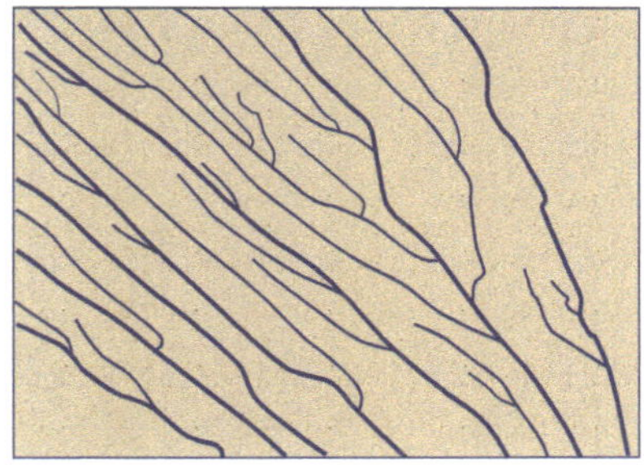

Figure 3.3 Parallel pattern. Adapted from von Bandat, 1962 and Strandberg, 1967.

inclined from the large stream to the base of the mesa. The low stream density and bright tones in the stream beds indicate a source of sandy sediment below the surface and in the vicinity of the mesa. Darker tones in the middle of the photograph suggest smaller grain sizes (silt or clay) exist at the surface.

These observations would lead one to conclude that no dangers exist on the ground here if there weren't several obvious areas of departure from dendritic drainage. I have marked four such areas with arrows. Whenever you find such departures, *never* ignore them; you must pay close attention to them. Where you find four anomalies, you will likely find more. If you can't explain the anomalous stream behavior using only the photographs, you are obligated to note them for inclusion in your list of areas to visit in the field.

So what are the four anomalies? Look closely at Figure 3.1, and at how sweeping and treelike the dendritic pattern looks. Now look at Figure 3.4 and at how linear the marked streams are. I have marked them and several more with dashed red lines. You can probably spot even more. These could be faults (that mesa we discussed earlier is higher in elevation for some good reason), or they could be joints

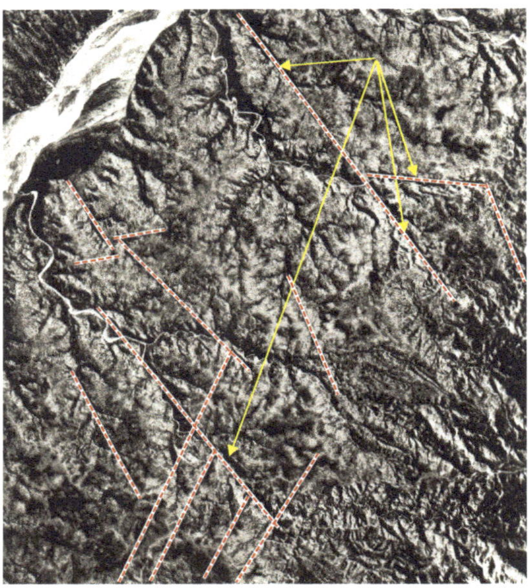

Figure 3.4 Panchromatic vertical aerial photograph of west Texas.

associated with underlying sandstone. The linearity doesn't mean there is danger; it does mean that you are obligated to explain it. Make an interpretation that includes several possibilities, choose specific areas that will help you rank the possibilities, and then visit these areas in the field. If your field visit proves your favorite interpretation wrong, you can always switch to one of the other possibilities. It is all right to be *wrong*—it is **not** all right to miss or ignore the anomalous streams and be *surprised* later when something goes wrong.

The linear departures from a dendritic pattern almost always indicate geologic structure of some kind. It is also possible that cultural activity such as trails worn into soft sediment could have been preferentially enhanced by stream erosion, but in this case that possibility would have a low ranking on my list. One could make the argument that the linear departures invalidate the assignment of a dendritic pattern. In this case, I would argue that the norm is dendritic, and the anomalous streams are, indeed, departures from that norm. If, however, there was a more predominant linearity or angularity to the stream pattern, a case could be made for an *angular* or *rectangular* drainage pattern.

3.1.3 Angular (Rectangular) Pattern

Figure 3.5 displays a panchromatic aerial photograph on which the drainage is controlled by rock joints, fissures, and faults. On Figure 3.5 I have traced the major drainages. Note the differences between this angular drainage and the free-flowing lines characteristic of dendritic patterns. Here the angularity forms the norm. Whenever large branches of drainage show angularity it is almost invariably indicative of rock structure. Other causes can be due to cultural activity. The stronger the angular imprint, the thinner the soil cover.

3.1.4 Trellis Pattern

Trellis patterns are formed in areas of folded sedimentary rock (anticline–syncline) where streams preferentially erode soft rock to form valleys, leaving harder rock to form ridges. Ridges are frequently cut by water gaps that give the trellis its characteristic shape (Figure 3.6). Drainage systems that closely follow geologic structure have formed on that surface. Conversely, patterns that obviously do not correspond to the geologic structure (a dendritic pattern over folded sedimentary rocks,

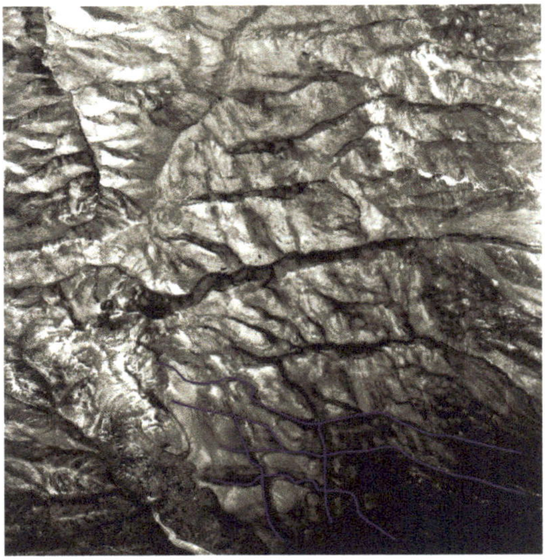

Figure 3.5 Panchromatic aerial photograph near Cortez Nevada showing angular drainage.

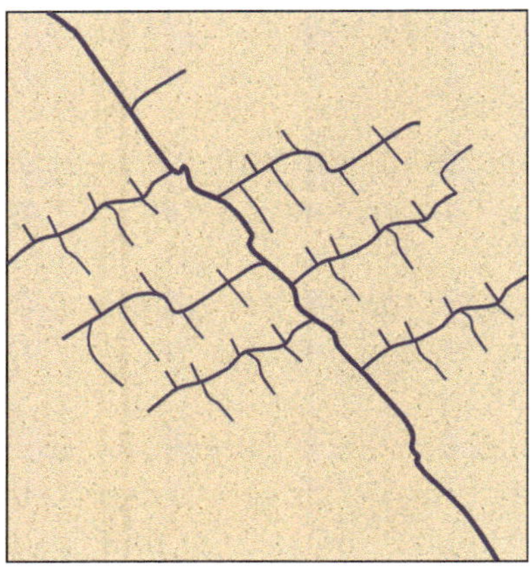

Figure 3.6 Trellis drainage pattern.

for example), were formed on another surface and superimposed on the visible geologic surface (an antecedent stream, for example).

Note the asymmetry of the stream lengths on the lower side of the secondary streams relative to the upper side. This asymmetry indicates that the geologic units dip toward the top of the figure. Always look closely at stream symmetry, especially if you believe that you have a

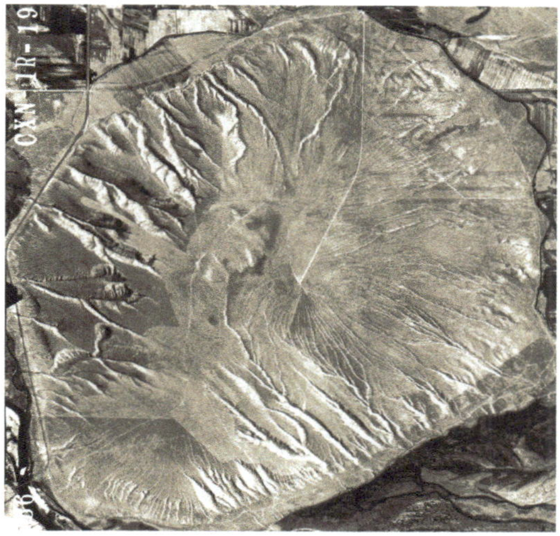

Figure 3.7 Radial centrifugal drainage pattern.

dendritic pattern. Asymmetry on the limbs of an apparent dendritic pattern indicates dipping strata, or strata that have been tilted due to faulting.

3.1.5 Radial Pattern

The radial pattern resembles spokes on a wheel and may be either centrifugal or centripetal. A centrifugal pattern (Figure 3.7) occurs when streams flow radially outward from a symmetrical hill, such as a dome or a volcano. A centripetal pattern develops where streams flow radially inward toward a basin or depression. A volcanic cone may have centrifugal drainage on its sides and centripetal drainage within its crater. Figure 3.7 is a rather obvious example, but be on the lookout for much more subtle radial tendencies which could form over active salt diapirs or active domes due to compressive stresses related to fault activity.

Streams will flow radially off of a dome, of course, but the formation of a dome will also obstruct regional stream flow. A regional dendritic stream pattern, for example, may display sweeping curves as the streams adjust to rising ground, forming the beginning of an *annular* pattern.

3.1.6 Annular Pattern

The annular patterns are formed when streams adjust to follow a circular path around the base of resistant hills. An annular pattern may

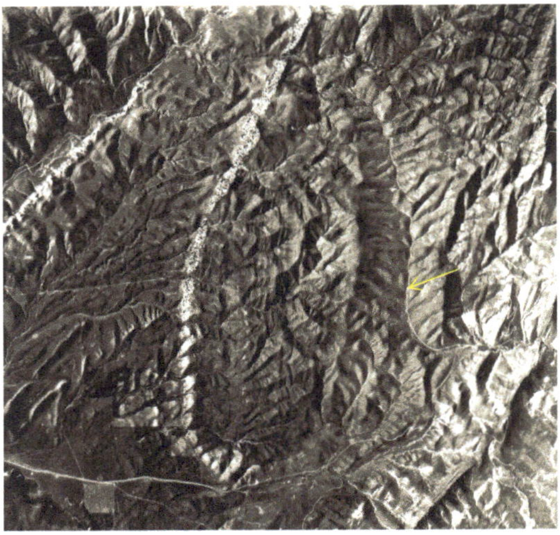

Figure 3.8 Vertical aerial photograph near San Luis Obispo, California showing annular drainage formed around a central core of resistant rock (arrow).

also develop as a modification of the radial pattern when an intruded body has up-warped bedded sedimentary rock of different strengths.

Note on Figure 3.8 how the stream (arrow) bends around a core of resistant rock. There is also a radial pattern associated with the central core. In this photograph, ringlike tributaries follow less resistant layers of tilted beds, intersecting radial channels at approximate right angles.

There are other stream patterns, but these are probably the most commonly encountered. I will cover one more, not because it is often seen, but because of the dangers it reveals. This is the *sinkhole* pattern.

3.1.7 Sinkhole Pattern

Figure 3.9 is an example of a sinkhole pattern. The sinkhole or swallow-hole pattern comprises short streams that end in depressions or that disappear and flow underground. These subterranean streams may re-emerge elsewhere at the surface as large springs. The bedrock underlying the areas in which this pattern develops is normally massive limestone in which pits and sinks have formed by chemical solution or by the collapse of caves (karst topography).

I have tried to convey in Figure 3.9 that most limestone develops a recognizable drainage pattern. The overall pattern will usually be

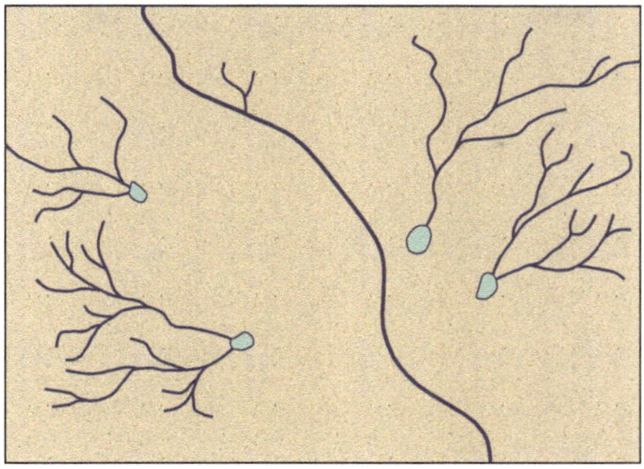

Figure 3.9 Sinkhole or swallow-hole pattern.

angular, because solution occurs preferentially along faults or fractures in the limestone. Commonly sinkholes will develop more readily along faults or fractures where differential movement has pulverized the crystalline limestone. As a result, alignments of sinkholes may result, which aids in delineating rock mass weakness zones. In Figure 3.9, I have imprinted a dendritic pattern over the sinkholes.

Unless you can prove otherwise (a difficult task, especially during litigation) always assume that ALL LIMESTONE IS CAVERNOUS. Cavernous limestone presents severe dangers to engineering works. It doesn't mean you can't build on cavernous limestone, but it does imply a large amount of detailed exploration will be required.

3.2 STREAM DENSITY

The density of streams in an area is related to the grain size and hardness of materials over which the streams flow. Coarse-grained materials are permeable and resist erosion, hence water tends to move readily through the body of the rock mass causing a low drainage density (i.e., few streams per unit area). Fine-grained materials are less permeable and more susceptible to erosion. Water, therefore, tends to be forced to flow along the surface, eroding a high drainage density.

It is sometimes worthwhile to quantify the stream network in an area, even if it is simply to be able to talk about it. There are numerous approaches to doing this. One of the earliest is known as the Horton–Strahler link numbering system[3,4] shown in Figure 3.10.

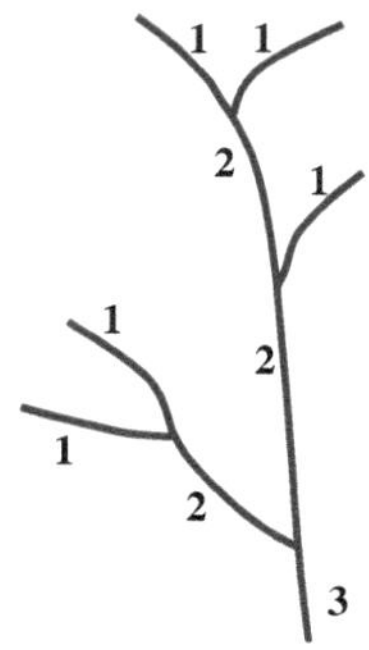

Figure 3.10 Horton–Strahler link numbering system.

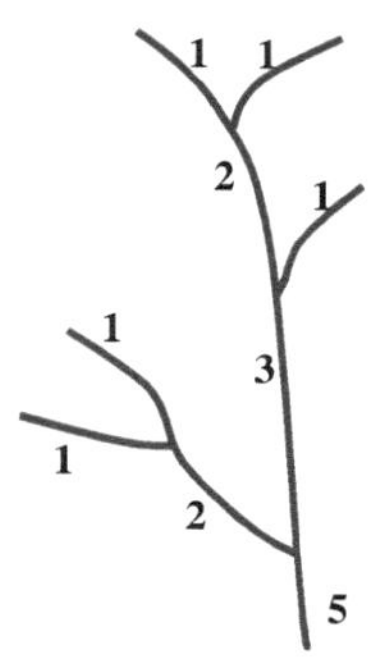

Figure 3.11 Consistent stream ordering system.

The Horton–Strahler system of link numbering starts from the smallest first-order streams and works down to the main stream. The numbering rule is that if two streams having the same number join, the rank is increased by one. If two streams having different numbers join, the rank is that of the larger of the two. Problems arise in using this system for estimating total outflow. A later approach called "consistent" stream ordering was first suggested by Schfeidegger[5] and later extended to stream magnitude by Shreve[6]. This network is shown in Figure 3.11. It shows channel links ordered by magnitude. The link magnitude is the number of sources contributing to any link. Here the channel formed by two source channels is ranked as the sum of the two.

The *magnitude* of the network is the number of sources (streams of order 1) for the network. So the network in Figure 3.11 has a magnitude of 5.

The network diameter is the maximum link distance in the network and is a measure of the longitudinal extent of the network, with mainstream length as its geometric analog. The diameter of the above network is 4.

The drainage density can be defined as

$$D_\mathrm{d} = \frac{\sum L}{A_\mathrm{d}}$$

where the numerator is the sum of the channel lengths in a basin having area A_d. This can also be approximated using the average distance from the end of the first-order streams to the top of the divide, l_s. So

$$\frac{1}{l_\mathrm{s}} = \frac{M}{\sum l} = \frac{\sum L}{A_\mathrm{d}}$$

where M is the magnitude of the drainage net (number of source streams), l is the length from each source stream (order 1) to the top of the divide, and the other terms are defined above.

These network parameters are regional parameters in that we need to be able to see the entire drainage net to use them. Obviously, if we have a large-scale photograph in which we can see only a small portion of a single gully, we can say little about the pattern or density. All is not lost, however, for there is much to be learned from a detailed look at gullies.

3.3 GULLY ANALYSIS

The following is a summary of gully characteristics for different kinds of geologic materials.

3.3.1 Granular and Resistant Materials (Sandstone, Carbonate, and Granitic Rocks)

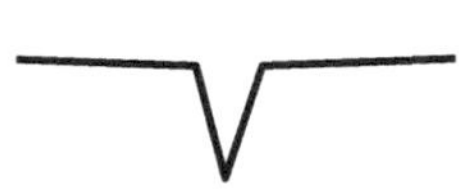

Narrow "V"-shaped gullies are found in sandstone, carbonate, and granitics. Streams in these materials are commonly short in length, have steep gradients, display sharp angular features, and contain sand and gravel in the stream bed.

Table 3.1 Channel Size and Shape Characteristics

Channel Size	Channel Shape
Width (w)	Width to depth ratio (w/d)
Mean depth (d)	d_{max}/d
Cross-sectional area (A = wd)	Asymmetry measures
Wetted perimeter (P)	$A^* = (A_r - A_l)/A$
Hydraulic radius (R = A/P ~ d)	$A_2 = 2x(d_{max} - d)/A$
Maximum depth (d_{max})	
Bed width (w_b)	
	A_r and A_l = cross-sectional areas to the right and left of centerline and x is the horizontal distance from the channel centerline to the point or centroid of the maximum depth.

3.3.2 Flaky and Nonresistant Materials (Shale and Silt)

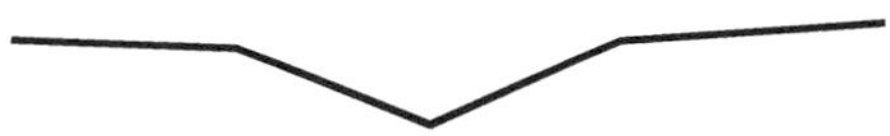

Gentle "V"-shaped gullies are found in shale and silt deposits. Streams in these materials are commonly long, have shallow gradients, display subdued features, and lack detritus in the stream bed.

3.3.3 Loess

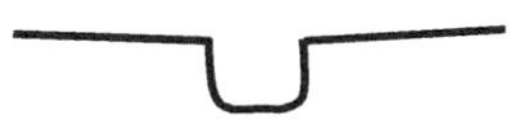

Loess (glacial, wind-blown material) is characterized by gullies having steep "U" shapes with shallow gradients and a lack of detritus in the stream bed. You can see these on streams adjacent to the Mississippi River in Kentucky.

3.3.4 Cross-Sectional Channel Forms

There are a number of metrics that are useful in describing the characteristics of a stream channel. Some of these are presented in Table 3.1.

REFERENCES

1. von Bandat HF. *Aerogeology*. Houston, TX: Gulf Publishing Co. 1962; p. 350.
2. Strandbert CH. *Aerial discovery manuel.* London: John Wiley & Sons 1967; p. 249.

3. Horton RE. Erosional development of streams and their discharge basins; hydrophysical approach to quantitative morphology. *GSA Bull* 1945;**56**:275–370.

4. Strahler AN. Quantitative analysis of watershed geomorphology. *AGU Trans* 1957;**38**:913–20.

5. Schfeidegger AE. Statistical description of river networks. *Water Resour Res* 1966;**2**(4):785–90.

6. Shreve RL. Infinite topologically random channel networks. *J Geol* 1967;**75**(2):178–86.

CHAPTER 4

Color Information

We have so far considered only textural information. Textural information is important because streams and cultural activity are well represented by textural patterns. To textural information, we will now add color, or brightness (amplitude) information. Think back for a moment about our previous discussion of the importance of photographic color to our ability to "read" aerial photographs. We touched briefly at that time on the interactions that can occur when sunlight strikes the earth's surface. Only three things can happen to the sunlight. It can be transmitted into the surface; it can be absorbed into the surface, in which case it is usually reemitted at a longer wavelength, usually as heat; it can be reflected back away from the surface, in which case we can record it with our sensor. The amount of sunlight reflected by the surface in the various visible wavelengths is a measure of the surface's *spectral reflectance* and is responsible for the color or brightness of a particular surface on an aerial photograph.

If you were to measure the brightness of a surface (basically its reflectance) in several different portions of the visible spectrum, you would find that the brightness changes as you move from the short wavelengths (blue) to longer wavelengths (red). It is this change in brightness in different portions of the spectrum that gives a surface its color. A surface that reflects more strongly in the green (0.55 μm) than elsewhere will appear green on an aerial photograph. A surface that reflects strongly in all colors will appear white to the human eye and on an aerial photograph. A surface with little reflection in all colors will appear dark.

If we were to plot the brightness of a particular surface on the vertical axis of a Cartesian coordinate system, and the color (wavelength) on the horizontal axis, we would produce a *spectral reflectance plot*. Spectral reflectance plots are convenient ways to compare the color (or spectral) characteristics of different materials. We will use them to illustrate the photographic colors that we can expect to find on photographs.

The spectral reflectance plot in Figure 4.1 shows the spectral reflectance characteristics of some common sedimentary rocks. Spectral reflectance curves are formed by crushing rock samples into powder and measuring the sample brightness using band-pass filters that allow only a limited wavelength band (color) through. To improve calibration, the brightness of the sample is compared with the brightness of some standard such as magnesium oxide (MgO), a white powder. The sample brightness in a specified spectral band is divided by the MgO brightness in the same band, and a relative reflectance value is presented.

What would something colorful look like on a plot like that in Figure 4.1? An imaginary spectral reflectance curve for a bright green shirt is superimposed on the figure. The reflectance would be low at 0.3–0.4 μm indicating high absorption of sunlight at those short wavelengths. Reflectance would soar to perhaps 80% at 0.5–0.55 μm, then plummet to near zero again into the longer wavelengths at 0.6–0.7 μm where absorption again dominates. Compare the spectral reflectance curve of the green shirt with the spectral reflectance of common sedimentary rocks. The rocks aren't very colorful, are they?

If you don't believe the spectral reflectance curves, take a look outside at the rocks and soil around you. Do you see a spectacular vista of bright colors? No, you see brown—beige if you will—and, as I live in Arizona, a little green perhaps. But even the greenest plants are rather dark. How lucky we are; God went out and painted the world beige.

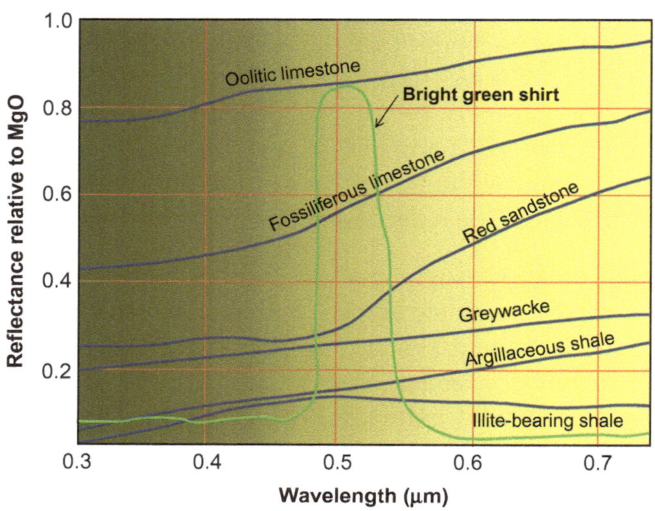

Figure 4.1 Spectral reflectance curves for six common sedimentary rocks (blue).

When we do see color, such as the beautiful reds of Canyon DeChelly in northeastern Arizona, it is related to iron staining or weathering products, which often have little to do with the underlying mineralogy. For this reason we can not expect to make unique mineralogic identifications on aerial photographs alone. Don't despair, though, we still can get valuable information about rock types from our photographs, and rock type has an important influence on dangers and natural hazards on the ground.

4.1 SEDIMENTARY ROCKS

We see from Figure 4.1 that sedimentary rock reflectances are relatively flat across the visible spectrum, indicating little color. The red sandstone appears so because iron ions absorb energy in the blue and green wavelengths (note the suppression of the reflectance from 0.3 to 0.5 μm) causing the spectra to increase in the red region. For all of the rock types, the reflectances increase toward the red, hence the beige color mentioned earlier.

Even without color, though, note the rather remarkable difference in overall brightness between limestone and shale. These brightness variations are dependable even at aerial photograph and field scales, and thus provide valuable information. With the above observations in mind, I present below some interpretive information for sedimentary rocks.

4.1.1 Sandstone

Resistance to weathering—Well indurated (hard) sandstone forms ridges and produces steep slopes and sharp landforms, including narrow "V"-shaped gullies in cross section that have short, steep gradients. Weathering of the jointed rock mass often produces blocky appearances with thick sand accumulation in gullies.

Drainage—Most sandstone is permeable, resulting in significant subsurface fluid flow. Even when well indurated, sandstone produces *coarse* drainage densities. If you trip over a gully with every other step in the field, you're not walking on sandstone. High permeability in sandstone allows water to quickly escape the roots of plants, so, in general, sandstone is not an efficient supporter of dense vegetation. Drainage patterns that develop on sandstone tend to be angular due to structural control.

Color—Pure silica sandstone will appear bright on an aerial photograph, easily as bright as reef limestone. Be careful, though, because there are often impurities in sandstone that may lower the reflectance. Iron staining is the most common darkening (reddening) agent. Here in the Southwest, reddish cliff formers are frequently sandstone units. Lithic contamination of silica sand will also darken sandstone, to the point where California coast graywacke (Figure 4.1) is nearly as dark as shale.

Structure—Sandstone commonly forms massive tabular beds. A common response of this geometry to tectonic forces involves the development of fracture patterns that produce a regular, measurable fracture spacing and identifiable sets. Fracture sets, although not necessarily orthogonal, commonly form at high angles to one another.

4.1.2 Conglomerate

Conglomerate shares many of the same characteristics with sandstone but is usually more limited in areal extent. Because of the large grains, conglomerate develops a drainage density that is even coarser than sandstone, and the large grains may contribute subtle textural differences due to shadows. At the upper edge of a bluff, for example, larger pebbles may migrate down slope, causing a light halo along the bluff face.

4.1.3 Shale, Siltstone, and Mudstone

Resistance to weathering—Shale, siltstone, and mudstone units are valley formers, producing gentle slopes and subdued landforms, including broad "V"-shaped gullies having long, gentle stream gradients.

Drainage—Shale, siltstone, and mudstone units have relatively low permeability, resulting in significant surface flow. This produces a high drainage density. The low permeability of shale and clay keeps water in the vicinity of plant roots, so, in general, shale and clay will support dense vegetation. This often obscures the surface of shale and further lowers the reflectance.

Color—Shale is commonly dark because of the lithic content, dark clay minerals, and carbonaceous matter, which is frequently a constituent of shale. The dark color of shale is usually indicative of marine origin. Marine shale dominates in frequency, areal distribution, and thickness. Fluvial and lacustrine shale have higher reflectances and often contain more silt. Siltstone, an abundant sedimentary rock type that is probably more prevalent on the landscape than shale, is often misidentified and labeled as shale by engineers. The mechanical properties of siltstone make siltstone an important class from an engineering/rock strength standpoint as well.

Structure—Shale can form thick, dense deposits, but shale response to tectonic forces is quite different from sandstone. Shale responds to deformation by forming numerous lenticular shear planes having separations measured in millimeters to centimeters. Hence, shale tends to be weak, requiring support for bearing loads. In addition, depending upon the type of clay minerals present, shale may present shrink/swell problems (smectite and chlorite minerals), or squeeze/deformation problems (kaolinite and chlorite minerals).

4.1.4 Carbonates

Carbonates (limestone and dolomite) share many of the characteristics of sandstone.

Resistance to weathering—Because of their crystalline nature, carbonates are ridge formers that produce steep slopes and sharp forms, particularly in arid regions. Dolomite is more brittle than the more plastic limestone. It often contains a matrix of sand and algae which forms thinner beds (flaggy) relative to limestone. Where strata are flat-lying, dolomite is often laterally associated and "seaward" of evaporates and sandy deflation flat facies (sabhka), which erode faster than dolomite. Limestone is often formed shoreward of dolomite. Even in humid regions, though, where significant solution has occurred, crystalline limestone may form precipitous cliffs and pillars. Limestone's relatively deep depositional environment often produces close spatial relationships with clay and shale as well as lime/clay mixtures within limestone units. Hence, one should expect most of the weathering products of limestone to be clay rich. A common derivative of limestone weathering is a fat, red clay called terra rosa, which can be seen in large parts of Indiana and Kentucky.

Drainage—The permeability of unfractured crystalline carbonate is low, but the hardness produces a coarse drainage density similar to sandstone. In places where solution has occurred, of course, subsurface flow may increase dramatically producing sinkhole drainage patterns. Limestone drains well, and most limestone produces angular drainage patterns. Watch closely for the sinkholes as a major indicator of limestone.

Structure—Like sandstone, limestone may form massive tabular beds that respond to tectonic forces in a similar manner.

4.2 IGNEOUS ROCKS

As discussed in the earlier section, igneous rocks are characterized by relatively featureless reflectance spectra as shown in Figure 4.2. As the composition of igneous rocks progresses from acidic (granite) to basic (dunite), however, the increasing iron content decreases reflectance; hence, basic igneous rocks appear dark on aerial photographs.

When discussing the appearance of igneous rocks on aerial photographs, it is convenient to focus our attention on two separate categories, namely:

1. intrusive igneous rocks
   plutonic
   hypabyssal
2. extrusive igneous rocks.

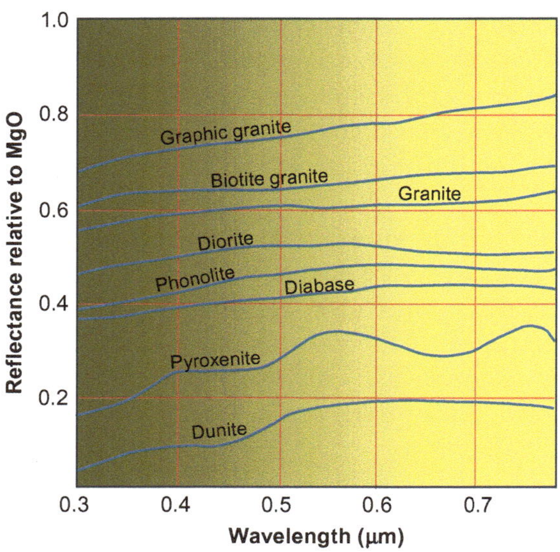

Figure 4.2 Spectral reflectance curves for eight common igneous rocks.

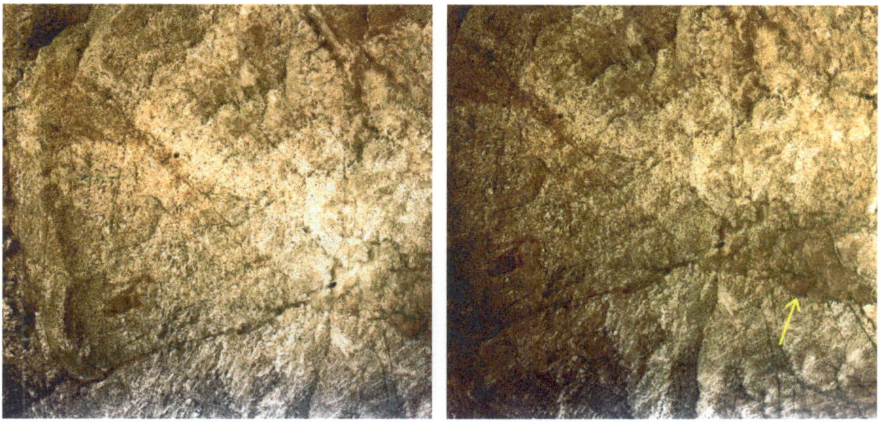

Figure 4.3 This vertical stereoscopic aerial photo pair shows granitic rocks forming the footwall of a metamorphic core complex in Arizona. Note the light photographic tone and the coarse, angular drainage controlled by faults and fractures. At the lower right hand edge (arrow) is a large graben, the existence of which was previously considered improbable within a footwall of a metamorphic core complex.

4.2.1 Intrusive Igneous Rocks

Intrusive igneous rocks are formed when magma crystallizes within the earth's crust. These intrusions are commonly divided into major intrusions, or *plutonic intrusions*, and minor intrusions, or *hypabyssal intrusions* (dikes, sills, etc.) which solidify within a transitional crustal depth from deep plutonic depths to the earth's surface. Approximately 95% of all plutonic intrusions have granite or granodiorite composition, hence they tend to display a relatively high reflectance, as shown in Figure 4.2.

Major intrusions include batholiths and stocks. Both can be large in size and are composed predominantly of granite and granodiorite. These are massive features that lack bedding. Because of their predominantly granitic composition they tend to be light in color. Texture is due to fractures and faults of the pluton caused by tectonic, cooling, and unloading stresses. Orientations of these discontinuities tend to be less regular than what we observe in tabular deposits. In arid regions expect angular and rectangular drainage patterns with coarse density due to the resistant crystalline nature of the rock (Figure 4.3).

Granitic rocks frequently weather into spheroidal shapes. Spheroidal weathering produces rounded or hummocky landforms, which cause low-order streams to form pincer shapes as they curve around the rounded landforms.

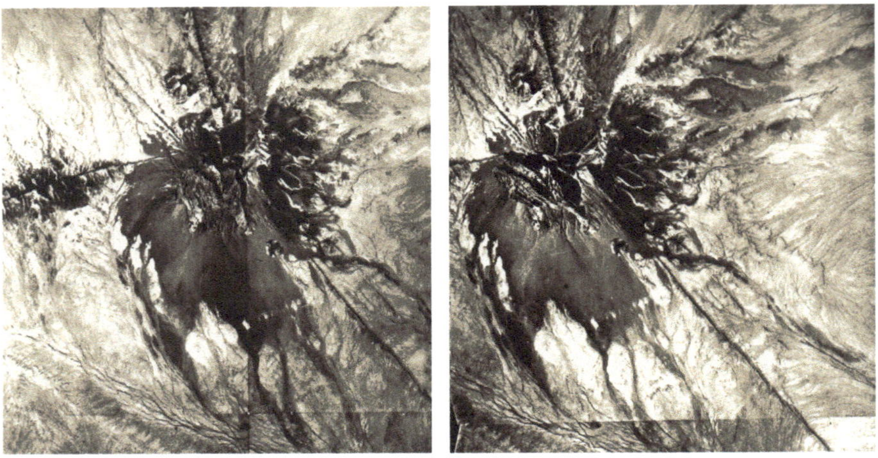

Figure 4.4 These vertical stereoscopic aerial photographs show hypabyssal rocks at Ship Rock, New Mexico. Note the long, straight mafic (dark) dikes extending to the upper left and lower right corners of the photographs.

Dark mafic intrusions are rare and tend to be limited in areal extent. Cooling and crystallizing from magma first, these bodies tend to form limited bodies surrounded by larger rock masses having more silicic mineralogy.

Hypabyssal intrusive rocks frequently stand out as resistant ridge formers having variable composition. Dikes occur frequently along faults or fractures which provide a convenient path to the surface. Some vertical dikes may strike a straight course for many miles as shown in Figure 4.4.

4.2.2 Extrusive Igneous Rocks

Unlike intrusive igneous rocks, which are mostly acidic in composition, most extrusive igneous rocks are basaltic in composition. With a few exceptions, such as back-arc volcanism having voluminous flows of flood basalts (Colombia Plateau, Washington and the Deccan Traps, India are examples), most extrusive flows stay in the neighborhood of the volcano or cinder cone that formed them. Hence that distinguishable landform provides one of the most powerful interpretation clues for extrusive igneous rocks (Figure 4.5).

If you are unlucky enough to have photographs that miss the flow source, there are still geomorphic clues in the flow levees, lava tubes,

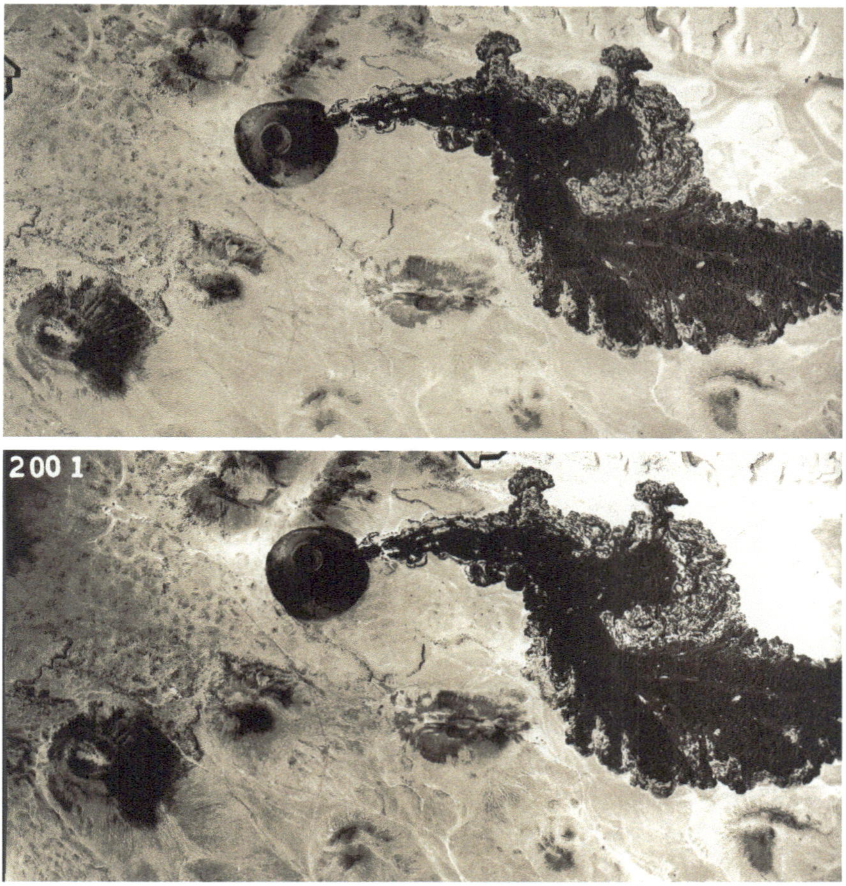

Figure 4.5 These vertical stereoscopic aerial photographs show cinder cones with associated basaltic lava flow, San Francisco Volcanic Field, Arizona. Arrange top photo on left.

and collapse zones, and flow ends to lead you to a correct interpretation. Other characteristics of extrusive flows are summarized below.

Drainage density—You don't see many streams in Figure 4.5, because the tops of extrusive flows are highly porous and permeable. Hence, they conduct subsurface water readily, forming in some local impressive aquifers even though the interior of the flows may be extremely dense and relatively impermeable.

Drainage pattern—Cooling cracks, faults, and fractures frequently result in angular or rectangular drainage patterns.

Dark photographic colors.

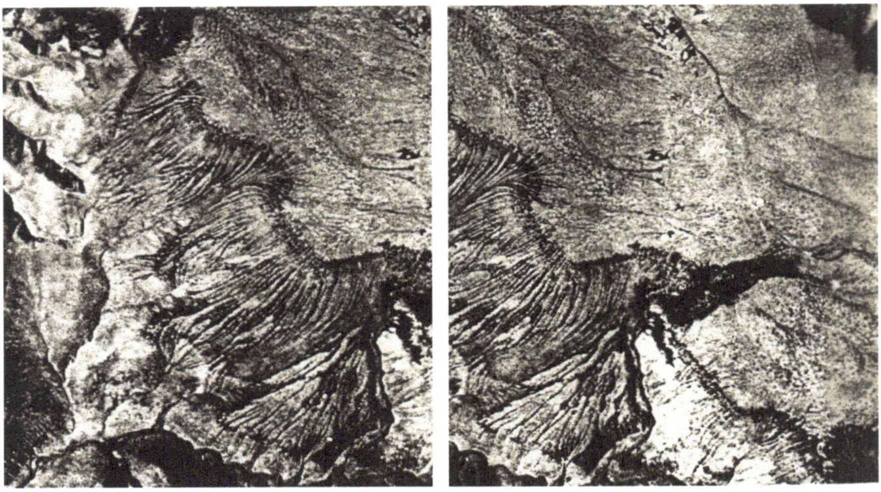

Figure 4.6 Stone stripes are a common occurrence caused by basalt blocks tumbling down from a basalt cap over light-colored silt.

Vegetation banding—Where extrusive flows intersect stream valleys we commonly see bands of vegetation growing preferentially along the valley sides where porous and permeable flow tops provide plants with ready access to water.

Stone stripes—Even in locations where basalts are covered by light material, they can frequently be detected by the unusual striped pattern formed by blocks of basalt falling from outcrops over lower, lighter material (Figure 4.6).

Mottled surface—The unusual surface texture seen on the basalt cap rock in Figure 4.6 is common to basalt flows. Here, it corresponds to grassy mounds of clay and silt, but in other cases, the vagaries of viscous flow can cause similar patterns and textures.

Basically, as a rule of thumb, if the source is not visible and flows appear thin, the flow was probably fluid in nature (pā hoe hoe). If the source is local and flow sides are steep and thick the flow was more viscous, similar to AA, or a flow having a more silicic chemistry.

Pyroclastic flows, unless welded, are generally thick, weak, and easily eroded. Hence, the drainage density tends to be high and the color light (Figure 4.7). In general, if it weathers like shale and drains

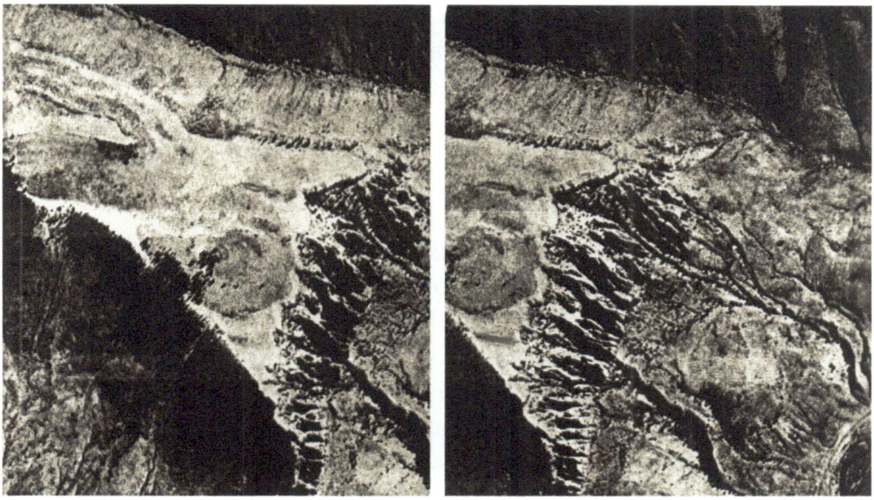

Figure 4.7 These vertical stereoscopic aerial photographs show tuff deposits in Colorado capped by welded tuff.

like shale but has brightness similar to sandstone, you might give some consideration to tuff.

Welded tuff is another story. It may have the brightness, hardness, and drainage characteristics of sandstone.

I have never seen a thick, massive deposit of welded tuff, though, nor the fracture characteristics that seem to accompany tabular sandstone units. Don't get me wrong, I have seen thick accumulations of tuff, but these are composed of alternating layers of welded tuff and airborne tuff (the Bishop tuff, near Bishop, California provides a good example) ultimately forming a thick sequence. My advice is, if you see alternating layers of hard and soft light-colored rocks, give some consideration to explosive ash deposits alternating with pyroclastic flows. The hard cap shown in Figure 4.7 is a welded tuff overlying soft airborne ash (tuff) deposits.

It is important to recognize these volcanic deposits, because, at least in the United States, thick accumulations of tuff and welded tuff indicate active, explosive volcanism. We see remarkably thick deposits of tuff in Arizona (Apache Leap Tuffs), Alaska (Aleutian Range volcanoes, Katmai, Redoubt, etc.), California and Nevada (Bishop Tuff and Cascade volcanoes), Oregon and Washington (Cascade volcanoes), and Wyoming, Idaho, and Montana (Yellowstone Caldera).

4.3 METAMORPHIC ROCKS

Metamorphic rocks are frequently difficult to interpret on aerial photographs because they are derived from preexisting rocks. The mineralogic, textural, and structural changes that occur during metamorphism can range from no visible change to a total reformation. Increasing temperature and/or pressure are the primary agents responsible for metamorphism. The *grade* of metamorphism refers to the temperature range under which metamorphism occurred.

Figure 4.8 displays visible reflectance spectra for some common metamorphic rocks. Note that as a group metamorphic rocks have a greater range of brightness and are more colorful than sedimentary and igneous rocks.

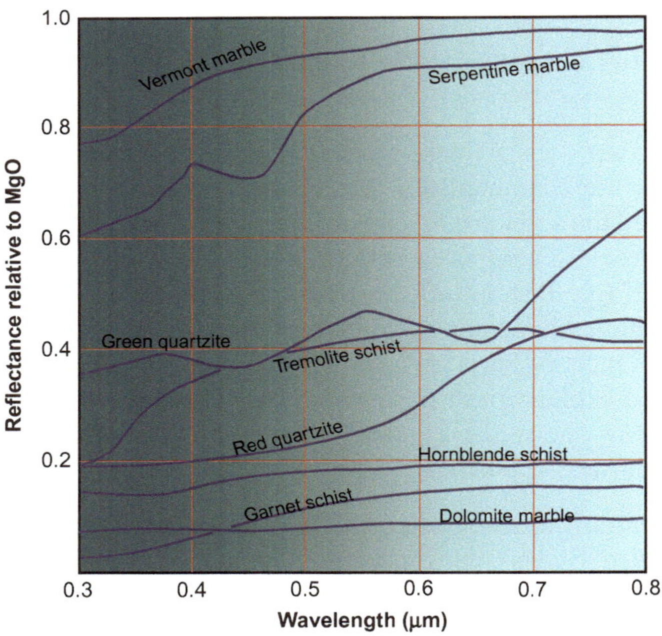

Figure 4.8 Spectral reflectance of eight common metamorphic rocks.

4.3.1 Schistosity Versus Foliation

Textures, as compared with color and brightness, provide the most dependable recognition keys for metamorphic rocks. Phyllosilicates are common minerals in metamorphic rocks of medium and low grades. Phyllosilicates found in metamorphic rocks include muscovite, biotite, chlorite, and to a lesser extent, talc, and serpentine. These flaky and tabular minerals are the minerals most responsible for the development of *schistosity*. Higher concentrations of phyllosilicates lead to more pronounced schistosity.

Foliation in metamorphic rocks is caused by contrasting mineral assemblages such as quartz–feldspar and mica–chlorite–amphibole. In intermediate- and low-grade metamorphic rocks, the foliation direction corresponds with the schistosity direction (if schistosity is present). Therefore, foliation is probably related to the same stress and strain regime that is responsible for schistosity in low- to medium-grade metamorphic rocks.

At high temperatures, though, the influence of stress decreases leading to a lack of schistosity in high-grade metamorphics. As a result, foliation becomes more significant as the grade of metamorphism increases. In addition, the phyllosilicates are replaced in high-grade metamorphism by nesosilicates (epidot, garnet, staurolite, kyanite, sillimanite, and andalusite), further reducing the tendency to produce schistosity. High-grade metamorphic rocks usually occur in close proximity to plutonic bodies, so be on the lookout for intrusive igneous rocks in the neighborhood.

In general, remember that the process of metamorphism frequently deforms the rock units. Look for:

- Vertical dips
- Sinuous or serrated ridges
- Kinks or kink folds
- Highly contrasting brightness and color.

4.3.2 Gneiss

Gneiss is a high-grade metamorphic rock having many of the aerial photograph characteristics of granite. In fact *primary gneiss* is largely indistinguishable from granite at aerial photograph scales. *Ortho gneiss* is essentially metamorphosed granite, so it too is similar in appearance to granite. Ortho gneiss, however, can develop foliation strong enough

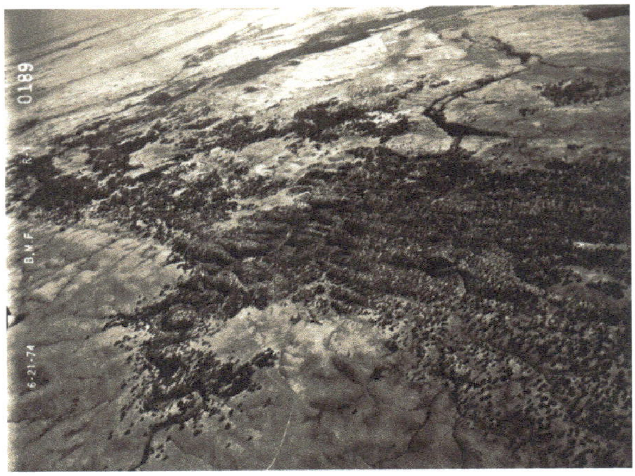

Figure 4.9 This oblique aerial photograph shows belted metasediment and metavolcanic rocks in the foothills of the Sierra Nevada batholith, California. Note the lack of predominant topographic highs or lows. The granitic core of the batholith is off the photograph to the upper right.

to be recognized on aerial photographs by subtle banding and elongation of drainage in a direction parallel to foliation. Depending on the grade of metamorphism, gneiss derived from granite usually provides equivalent strength properties. Some of the rocks in Figure 4.3 may, indeed, be gneiss.

Para gneiss is a metasediment (high-grade metamorphism of a sedimentary rock); hence, it may retain some of the morphological characteristics of the original sediment from which it developed. Look for sharp, crested ridges, and angular drainage patterns which tend toward parallel. Para gneiss is common to belted metamorphics around the cores of plutons. Figure 4.9 shows an oblique aerial photograph of belted metamorphics along the western foothills of the Sierra Nevada batholith in central California.

4.3.3 Schist

Schist is darker in color than gneiss. In color photographs or with the naked eyes, schist appears gray. In Figure 4.8, note that the overall reflectance of schist is lower than 50% that of the MgO standard. Schist has an abundance of phyllosilicate minerals, so schistosity and foliation are pronounced. Look for rough topography with banded, ribbed, or striped appearance. Drainage patterns are often trellis to angular and faults are common.

Schist units are highly inhomogeneous and anisotropic, possessing alternating hard resistant zones with weak crumbly ones. They are commonly landslide prone, especially where weathering has proceeded deeply into the unit. The existence of schist is a red flag for potentially dangerous ground conditions.

4.3.4 Slate

Slate derives from low-grade metamorphism of fine-grained, clay-rich sediment. The parent sediment is weak and tends to flow plastically when deformed. This behavior results in cleavage planes and recrystallization. Illite, sericite, and chlorite, the principal flaky minerals in slate, are commonly rotated into parallel (dispersed) alignment and recrystallized. Other minerals include quartz, feldspar, and hematite.

On aerial photographs slate is often multicolored and dark. Drainage patterns on slate are routinely angular, often rectangular and dense. Slate topography tends to display a wrinkled appearance having rounded hill tops and steep slopes. To compliment the wrinkled analogy, there commonly are few prominent hill tops or valleys; all topographic features have approximately the same height and depth.

Although individual plates of slate can be "ringing" hard, the overall performance of slate as a foundation rock mass is poor due to its highly inhomogeneous and anisotropic nature. Planes of weakness are numerous and slope stability is a problem due to weathering of the phyllosilicates. Expect squeezing and swelling problems associated with illite and chlorite.

4.3.5 Quartzite

Quartzite is due to metamorphism of sandstone. It forms prominent topography due to its hardness and resistance to erosion. Quartzite is also brittle, though, so talus and surfaces may look as if they have been shattered. Quartzite can provide strong foundations, but fractures are common, so rock slides and toppling failures are common also.

4.3.6 Marble

Regional metamorphism of carbonate rocks simply recrystallizes them forming either calcite or dolomite marble. Marble is common and possesses a sugary texture with bright overall reflectance. Marble can form massive rock masses, but the strength and endurance of marble is generally low.

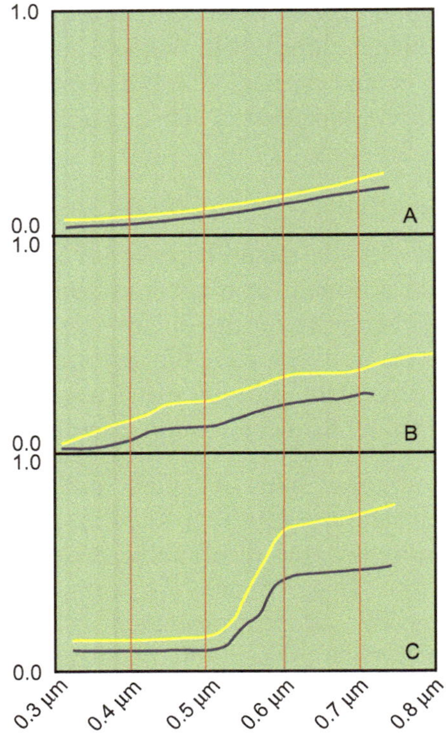

Figure 4.10 The figure shows reflectance spectra for three common soils: (A) grassland soil, (B) forest soil, and (C) common to tropical and southwestern US soils.

4.4 SOILS

The appearance of soils on aerial photographs is a function of a number of factors including:

Mineralogy—Silicon, feldspar, and gypsum tend to increase soil brightness while flattening the spectral reflectance.

Texture—Clay, because of its low permeability, tends to produce the highest degree of surface texture due to stream erosion, whereas sand, because of its high permeability, produces the lowest.

Humus, water, and iron oxide content—Humus, water, and iron oxide have the most profound effects on the appearance of soil on aerial photographs, but in quite different ways. Figure 4.10, for example, shows spectral reflectance plots for three common types of soils. Note the effect of water on the soil brightness. Wet soil spectra are shown in blue and dry soil spectra are shown in yellow.

In all cases, the addition of moisture to soil decreases brightness approximately the same amount, but note that water does *not* change the spectral shape (color). A moist soil will, therefore, appear darker than the same soil in a dry state, but the distinguishing color characteristics of the soil will remain the same.

The spectral shape of soil is controlled by the ratio of humus to iron oxide. The soil depicted on Figure 4.10A corresponds to a rich grassland environment in which the soil is rich in humus and carbonates, both of which contribute to suppressing and flattening the spectra.

Figure 4.10B corresponds to a soil of forest environments which are rich in humus, alumina, and iron oxides. The iron oxide produces some spectral variability due to iron absorption in the blue and green wavelengths (0.4–0.5 μm), but the effect is largely smothered by the effects of humus and alumina.

Figure 4.10C presents the spectra of soils rich in iron oxide. Note the strong iron absorption band in the blue to green region of the spectra. This iron absorption band is clear in these soils because of the lack of humus. We typically find these soils in humid regions (laterites) and in the soils of the arid Southwestern United States.

Although not indicative of mineralogy, subtle changes in the ratio of humus to iron oxide produce quite distinctive changes in soil appearance, which are particularly apparent on color photographs. If you are mapping soil units, as you might, for example, if you are concerned with soil swelling or collapse potential, go ahead and map soil units based on these subtle differences even though you may not know what the mineralogy of the soil units is. You can always assess mineralogy, soil development, age, swell and collapse potential, etc. later on the ground and in the laboratory.

Vegetation—The type and density of vegetation is also a valuable clue to distinguishing among soil units on aerial photographs.

4.5 STRUCTURAL PATTERNS

Structural patterns, or indications that a geologic structure varies from horizontal stratification, are reflected in topography and in the degree to which drainage patterns vary from dendritic.

Structural patterns of horizontal strata—Dendritic drainage patterns usually develop on horizontal strata and cut canyons or valleys in which progressively older rock units are exposed at depth (Figure 4.11).

As a result, the aerial photographic patterns of horizontal strata parallel stream valleys and produce a dendritic pattern on an aerial photograph. Although topographic contour lines are not shown on aerial photographs, the contacts of rock units of horizontal strata will parallel the contours. Escarpments and gentle slopes generally develop on resistant and nonresistant beds respectively and thus produce variations in width on vertical aerial photographs. On steep cliffs, the upper and lower contacts (as seen on vertical aerial photographs) may not be clearly seen; another reason to alternate field visits with aerial photograph analysis. On gentle slopes, even in the same formation, the contacts will appear farther apart. It is apparent, therefore, that the width of outcrops as seen on an aerial photograph is not dependent solely on the formation thickness. Gently dipping strata will develop the same basic outcrop pattern as horizontal beds. The contacts between rock units in gently dipping strata, when traced far enough up a valley will be seen to cross

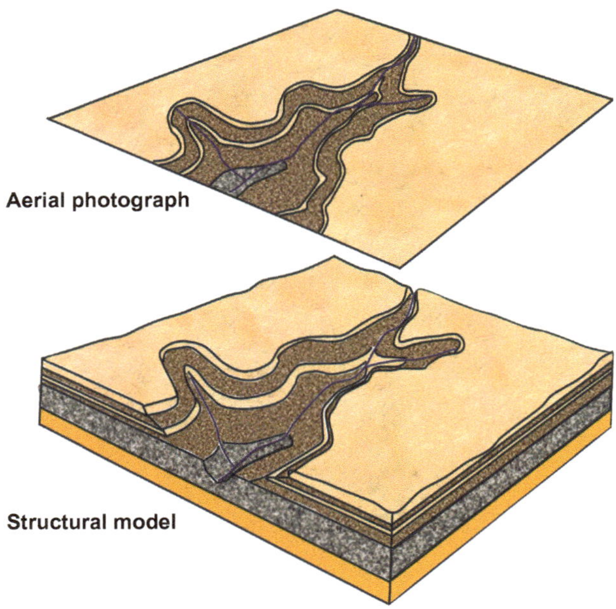

Figure 4.11 Structural patterns of horizontal strata.

topographic contours and to form a large V-shaped pattern which points in the direction in which the beds dip.

Which is the oldest bed and which is the youngest on Figure 4.12? How do you know this is true? Establish a working hypothesis to describe this relationship of bed sequence. Can you think of an instance in which this hypothesis would not be true? Is it equally true for igneous, sedimentary, and metamorphic rocks?

Structural patterns of inclined strata—When a sequence of rocks is tilted and truncated by erosion, the outcrop pattern will appear as bands which, on a regional basis, are roughly parallel.

Important variations in details of the basic pattern are developed in areas dissected by erosion and should be carefully analyzed, for they provide important information concerning the subsurface structure. When dipping strata are traced across a valley, a V-shaped outcrop pattern is formed which points in the direction of the dip. The size of the outcrop pattern V is inversely proportional to the magnitude of the dip.

Folding, which is the common cause of inclined strata, is one of the most common types of structural deformation, and is found in complex mountain ranges and in less deformed lowlands and plateaus. A convex upward fold in which the limbs dip away from the axis is referred to as an anticline. Anticlines have the oldest rocks in the core or center. Synclines are folds that are concave upward with

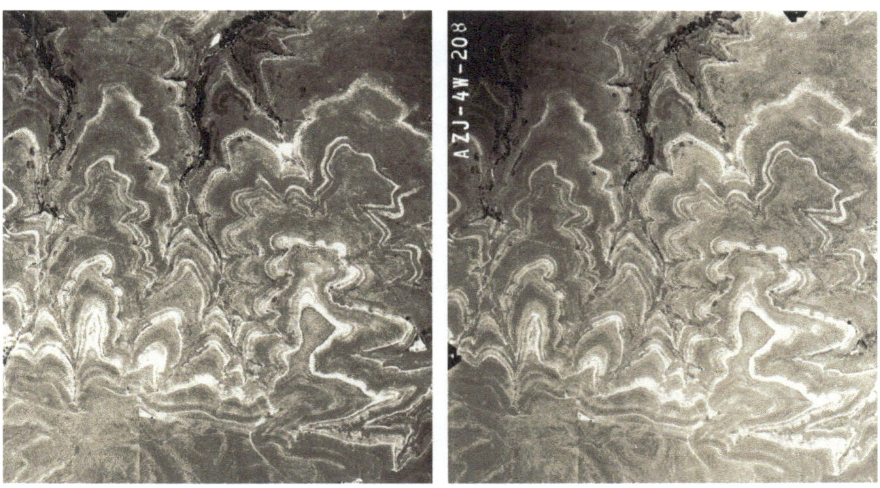

Figure 4.12 These vertical stereoscopic aerial photographs show outcrop patterns of horizontal marl beds (light tones) in Kansas.

limbs dipping toward the axis. Synclines have the youngest rocks in the center. In Figure 4.13, two anticlines flank a syncline.

In some folds the axis is horizontal, but more commonly it is inclined. The inclination of the axis is referred to as its *plunge* and is of importance in analyzing the 3-D aspects of the fold. Plunging folds that have been exposed by erosion form a characteristic zigzag outcrop pattern such as the one in Figure 4.13. A plunging anticline forms a V-shaped outcrop pattern with the apex (nose) pointing in the direction of plunge. Plunging synclines form a similar pattern, but the limbs of the fold open in the direction of plunge. Most of the Appalachian region, and especially eastern Pennsylvania, is composed of closely spaced, tightly folded anticlines and synclines.

Suppose now that Figure 4.13 was also folded gently about the axis shown in red into another anticline perpendicular to the original axes. This would produce a doubly plunging set of folds whose outcrop patterns would appear to be an elongated dome. Drainage patterns are helpful in interpreting domes because:

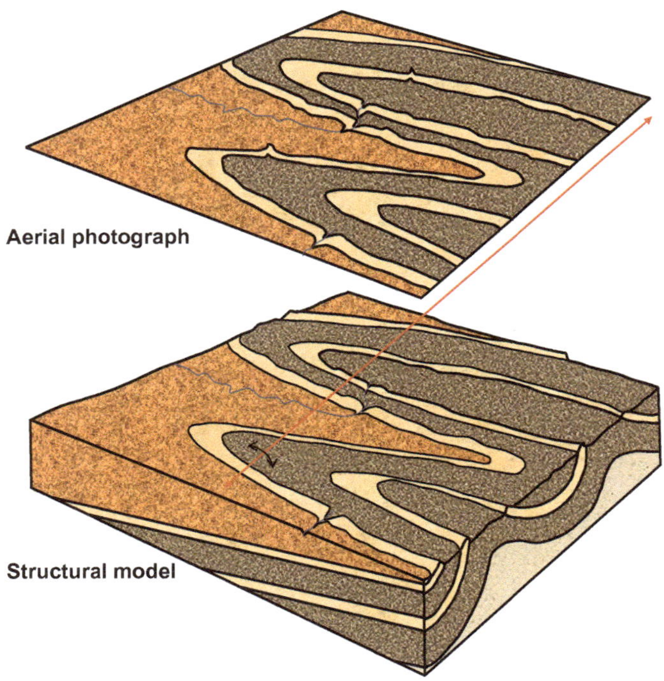

Figure 4.13 The figure displays structural patterns of inclined strata.

1. they tend to form circular patterns as streams develop on the less resistant beds, and
2. streams cutting across the resistant beds permit one to apply the rule of V's to interpret the direction of dip.

Structural patterns of faults—Fault patterns on aerial photographs are distinctive in that they abruptly offset structures and terminate contacts. Thrust faults (see point A in Figure 4.14) generally dip at low angles and have displacements commonly measured in kilometers. The pattern of the fault trace is usually irregular and similar in many respects to that produced by low-dipping, angular unconformities. The trace of the thrust fault commonly forms V's across valleys. These V's point in the direction of the thrust fault dip. Erosion may form windows (fensters) through a thin thrust sheet or produce isolated remnants (clippen) of the upper thrust sheet in front of the main thrust block.

Normal and reverse faults generally dip at relatively high angles, therefore, their outcrop patterns tend to be relatively straight (see point B on Figure 4.14). One caution should be considered in looking

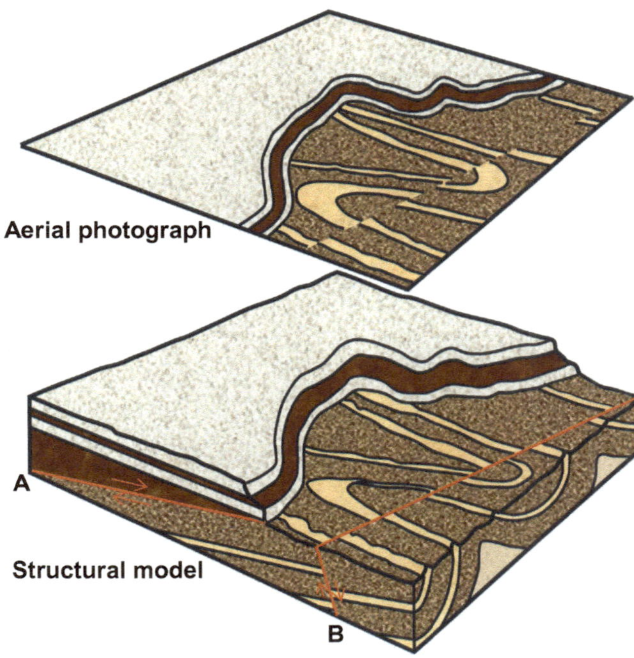

Figure 4.14 The figure shows some structural patterns of faults.

at Figure 4.14. Note that the offsets of the inclined strata by the normal fault, as seen on the aerial photograph, appear to be in the horizontal direction (similar to a strike-slip fault) rather than in the vertical direction. This occurs because erosion has caused the uplifted beds to migrate down dip. Recognition of the difference in apparent offset of the anticline versus the syncline will provide help in interpreting the sense of fault movement correctly. Just remember to not make snap judgments. Sometimes things are not as they seem.

My purpose in presenting these last two chapters is to illustrate commonly encountered surface conditions so that we may begin to recognize subtle variations that may indicate dangerous conditions. Figure 4.14 shows some of these variations involved when faults disrupt a folded sedimentary sequence. Can you see any indications of faulting in Figure 4.15? The presence of faults does not necessarily imply dangerous conditions, but it certainly serves as a red flag, warning us that dangerous conditions may be present, and that more study is warranted. Never ignore these red flags. We will see more of these in the following section.

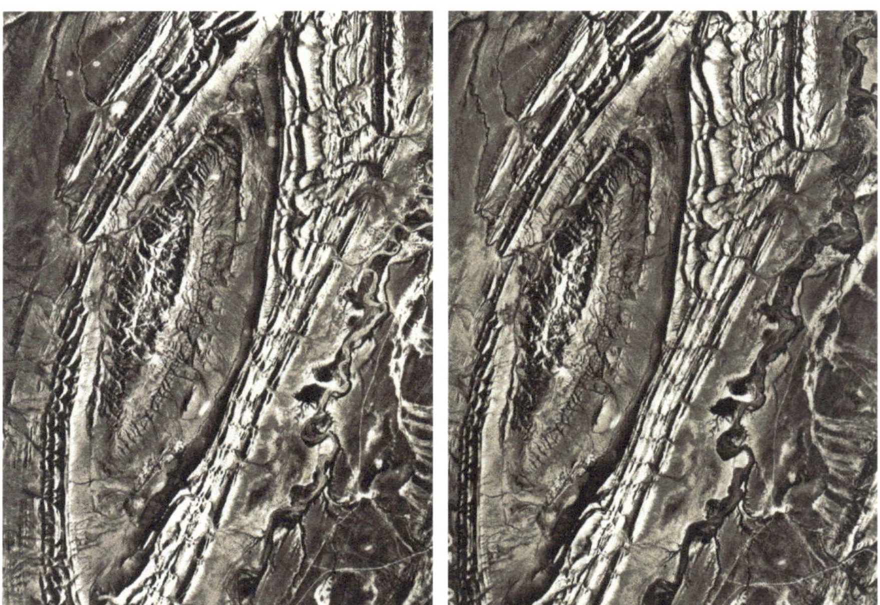

Figure 4.15 These vertical stereo aerial photographs show a doubly plunging anticline in Wyoming.

In the future, careful study of faults in most of the country or world may relate increasingly to water exploration, cultural development, and contaminant mitigation. Fault recognition on aerial photographs will no longer serve only as providing red flags for construction hazards, but will serve as an early stage for intense study of sustainability. As such the recognition, mapping, and analysis of fault geometry will become increasingly important.

PART II

Recognizing Hazards on the Ground

Now that we have established a baseline for recognizing natural hazards (i.e., departures from the norm), we are ready to see how some of the most dangerous hazards appear on aerial photographs. When we discuss a natural (not human caused) dangerous hazard, we are dealing with a natural condition that has a probability of producing some measurable impact on the human community. The impact of such a hazard, incidentally, is always a negative one. Some examples of dangerous conditions are:

- Earthquakes
- Landslides
- Floods
- Volcanic eruptions
- Ground subsidence
- Collapsing, squeezing, or swelling soils.

It seems clear that one will be hard pressed to arrive at reasonable estimates of the probability of future dangerous or hazardous events based solely on interpreting aerial photographs. It should seem just as clear, though, that the best way to avoid hazards is to stay away from areas where hazardous events have occurred in the past. I know it seems crazy, but if a particular geologic unit has experienced landslides in the past, it just might be possible that landslides will occur there again in the future—so one of the prerequisites to hazard avoidance is hazard recognition. This *can* be accomplished with the help of aerial photographs. Let's look at some of these hazardous geologic events individually and see how they can be recognized on aerial photographs.

This is a good time for a minor warning. I love it when aerial photographic observations can be used to derive quantitative information that scientists and engineers can use as an aid in design. In each of the following chapters, I have included some equations and mathematical models for that purpose. If you are not interested in these departures from my main mission of image interpretation, simply skip over them. Subsequent material will not rely on any of the models or equations.

CHAPTER 5

Dangers from Earthquakes and Faults

As we all know, earthquakes (tectonic ones, i.e., those caused by the movement of plates forming the earth's surface) occur when stored elastic energy is released by dislocation (rupture) through asperities along a fault. An asperity is a portion of the fault that is stronger than adjacent portions. Asperities can be caused by variations in fault geometry (curves), which create concentrations in stress along and within fault zones, variations in pore pressure, variations in rock type both within or to either side of the fault, variations in internal mineralogy, the degree and type of cementation, and/or gouge or altered material along the fault plane, variations in the presence or absence of interstitial or bound water in the fault zone or plane that determines the effective strength of a given rock unit or sedimentary unit, and other variations. Rupture of large asperities, cause large earthquakes known as *characteristic* earthquakes. Large earthquakes—the worst kind—threaten humans, their engineering works, and natural systems in the following ways.

- *Fault rupture*: Building a structure to withstand several feet of differential displacement through it can be done I suppose—but why would you want to try? It is best to avoid placing structures across active faults. In fact that should be our first rule: **do not build on an active fault** or even a so-called "inactive splay" within an active fault zone. Single faults are almost always a single splay within a larger or wider fault zone. Subtle splays are often overlooked. I frequently see such splays going unmapped or unnoted even on aerial photos. Fault systems and splays can be, in fact, dynamic with their activity varying over human and geologic time scales depending on the changes (if any) in orientation of the stress field or partitioning of stresses when asperities lock. In other words, an incipient or inactive splay can become active tomorrow if stresses increase or faults in other directions (antithetic Reidels) rupture. Such splays usually exhibit expressions of reactivation visible on aerial photographs. Other disruptions of natural systems due to faulting include changes in spring discharge, stream capture, base level drops or rises that produce ponds or lakes, marine incursion, accelerated erosion, vegetation die-offs, and habitat loss from marine incursions (vegetation die-offs in Portage, Alaska as a result of over 3 m crustal deformation during the Alaska earthquake of 1964 is an example).

- *Ground shaking*: Once active faults have been identified and avoided, strong ground shaking becomes a major concern. Regional observations and elastic/inelastic theory demonstrate that strong ground shaking attenuates (decreases) with distance from an earthquake source zone. Other considerations (resonance within soil and buildings), however, show that once active faults have been avoided, proximity to the active fault should not necessarily be an overriding concern. Photographs taken after the 1906 San Francisco earthquake, for example, demonstrate that poorly built structures on firm ground, even when located only a few feet from the San Andreas Fault, performed better than structures built on uncontrolled fill located several miles from the San Andreas Fault along the waterfront in San Francisco.
- *Landslides*: Landslides are a secondary effect of strong ground shaking. In fact, some earthquake specialists use the occurrence of landslides to gauge the severity of ground shaking in a given area. This is a mistake. Don't do it. Dynamic loading is but one of the factors affecting the stability of a slope. Others include:
  - Strength and frictional properties of materials at the slide plane.
  - Geometry of the slide plane, discontinuities, and slope.
  - Pore pressures at the slide plane.
  - Antecedent rainfall.
  - Loading conditions both static and dynamic.

  Dynamic loading conditions associated with earthquakes are clearly only a partial contributor to slope instability. Nevertheless, if slopes in the epicentral area of an earthquake are only marginally stable, strong earthquake ground motion can cause landslides.
- *Liquefaction*: Liquefaction occurs in loose, saturated granular (sand mostly) materials subjected to prolonged strong ground shaking. As the loose sand attempts to compact in response to earthquake vibrations, water in the pores can't quickly escape, so pore pressures begin to increase. As vibration continues, pore pressure continues to increase until the upward pressure of the water equals or exceeds the downward forces of gravity, resulting in sandy water rather than wet sand. As the sandy water can not sustain shear forces, any

structures built on the liquefying sand sustain damage. According to Casavant[1], on the Colorado Plateau, apparently competent and lithified sandstone or shale units overlie fluid-rich, less competent sand- and silt-rich units. In these cases, substrata liquefaction may result in fluid migration up fractures or new fracturing through overlying thin brittle units to deform surficial units and deposit sand laterally as thin, sandy deposits visible on aerial photographs.

- *Tsunami and seiches*: Tsunami are seismically generated sea waves. On December 26, 2004, for example, a magnitude of 9.1–9.3 earthquake off the western coast of Sumatra, Indonesian generated a tsunami in the Indian Ocean that exceeded 30 m in amplitude in some places and killed more than 225,000 people in 11 countries. Seiches are generated in fully or partially enclosed bodies of water such as lakes and bays when the frequency content of the seismic waves matches the resonant characteristics of the water body. Visualize the back and forth sloshing that can occur by shaking a tub of water.

All of the dangers mentioned above are created by large earthquakes resulting from the rupture of faults; it seems clear, therefore, that identifying faults capable of generating large earthquakes is the first step in avoiding earthquake damage. So let's take a closer look at some of the characteristics of active faults that we can see on aerial photographs.

I will be referring back to Chapter 4 several times in this chapter. You may wish to open Chapter 4 and tile it next to the current chapter for easy reference.

5.1 IDENTIFICATION OF FAULTS

Faults are dislocations within a geologic unit, or units, across which relative movement has occurred. Faults can be recognized because they produce discontinuities in color or texture. Discontinuities often occur as relatively abrupt linear or curvilinear elements or segments. The elements can be aligned, or be offset in a parallel or subparallel manner. Discontinuities in color or texture are caused by variations in one, or a combination, of the following:

- Lithology
- Vegetation
- Moisture

- Topography
- Erosion.

5.1.1 Contrasts in Lithology

Faults commonly bring different rock types into contact. You can appreciate this by referring back to Figure 4.15. Rock units having measurable differences in reflectance (color) may be readily distinguished on imagery even though it is not possible to identify precise mineralogy (Figure 5.1). Occasionally faults are even detectable within a single rock type, because of differences in erosion or weathering (texture) across the fault.

Diffuse fault zones and individual faults within are sometimes identified by textural changes in stream reach morphologies that create a "swarm or corridor" containing like-oriented linear reaches and/or orthogonal bends. Casavant[1] has found this to be the case in some alluvial covered bedrock settings. Little contrast in color or texture to either side of the zone is evident; rather, the alignment and matching of stream segments forming different washes provide the few clues to the presence of underlying fault zones (The North Slope of Alaska and the Rosemont area of the Santa Rita Mountains, Arizona provide two examples). Figure 5.2 shows how the texture and overall pattern of rock masses can be useful in identifying faults. The dominant textural component of rock masses seen using aerial photographs is the texture associated with stream patterns. Rock mass structure and hardness are both reflected in the pattern and density of the drainage system.

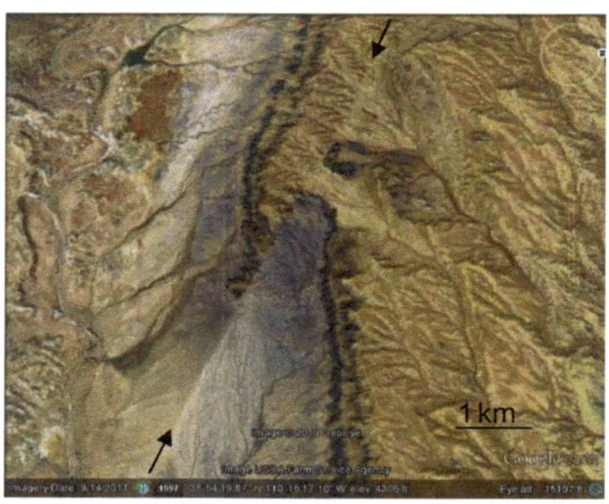

Figure 5.1 The Google Image shows a fault, identified by arrows, which produces a contrast in color due to lithology differences (Upper Cretaceous Mancos shale, dark, abutting Holocene sediment). The location is 13 km northwest of Green River, Utah.

In addition to stream patterns, variations in rock strength, hardness, and stress history result in fracture patterns having widely varying orientations and spatial frequencies. Changes in the pattern, density, or texture of streams and fracture systems often denote a fault contact, though ancillary data should always be used for verification.

Faults occurring within a single rock unit are often indistinguishable using color alone; other diagnostic interpretation keys must be used. The difficulty in locating faults within continuous rock units may profoundly affect a project. For example, I am aware of projects in which a fault was easily traced for tens of kilometers only to appear to stop abruptly at an igneous intrusion. It did not appear to displace the intrusion, but could be found and traced for tens of kilometers on both sides. Earlier, investigators had used the date of the intrusion to argue that the fault was no longer active. In two cases of which I am aware, active faults were classified as inactive based on radiometric ages of "unfaulted" igneous bodies. In both cases, careful interpretation of large scale, low-sun illumination aerial photographs to enhance fault features, coupled with fracture density mapping (see discussion of fracture tensors and fry diagrams in Chapter 6) and ground studies demonstrated that the fracture density increased near the projection of the fault through the igneous intrusion and that faults displaced the intrusions along wide zones of complex fractures. Given the importance of active faults to earthquake engineering, conclusions regarding the temporal frequency and activity of faults in such cases should not be made by relying on published literature or conventional maps or images

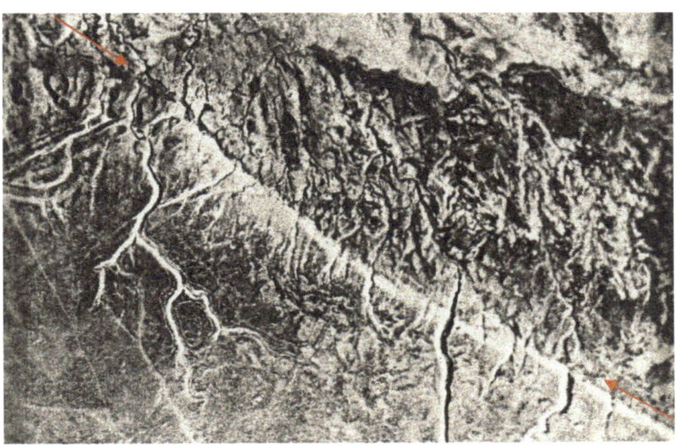

Figure 5.2 Fault, between arrows, identified primarily by texture contrast.

alone, but should rather be made only after a thorough aerial and ground analysis.

Conclusions—If a fault produces a variation in lithology, the parameters most likely to be useful for detecting these variations are:

- *Contrasting reflectance (color)*
- *Contrasting texture or pattern.*

5.1.2 Contrasts in Vegetation

Different rock or soil types may support different plant assemblages or may support a single plant assemblage but with varying growth characteristics. Therefore, even though the actual rock surface may be partially or totally obscured, differences in the type, age, vigor, density, or pattern of vegetation may provide clues to fault location.

Vegetation variations may occur as a color contrast across a fault, in which case the contrast is generally due to differences in the type of vegetation on one side as compared with the other. Vegetation contrasts may also occur as texture contrasts across a fault due to changes in vegetation type or more typically vegetation density. Moisture dammed or intercepted (pirated) by a fault may promote more dense vegetal growth on one side than the other. Faults also often serve as conduits for fluid flow, promoting vegetal growth along the fault and associated fractures (Figure 5.3).

In such cases, alignments of vegetation are formed, that may extend for tens of kilometers. Vegetation along fault scarps may be of a younger age than vegetation on either side yielding a relative age relationship for determining activity along the fault. Vegetation downstream of a fault is usually under stress. Stream networks may be altered with time due to infilling by wind deposition, lack of scouring by prefault flow regimes, erosion, etc.

Conclusions—The parameters most likely to be useful in detecting variations in vegetation are:

- Differences in reflectance of plants due to stunting or abnormal growth. These differences are particularly apparent in the near infrared portion of the spectrum (Chapter 10).

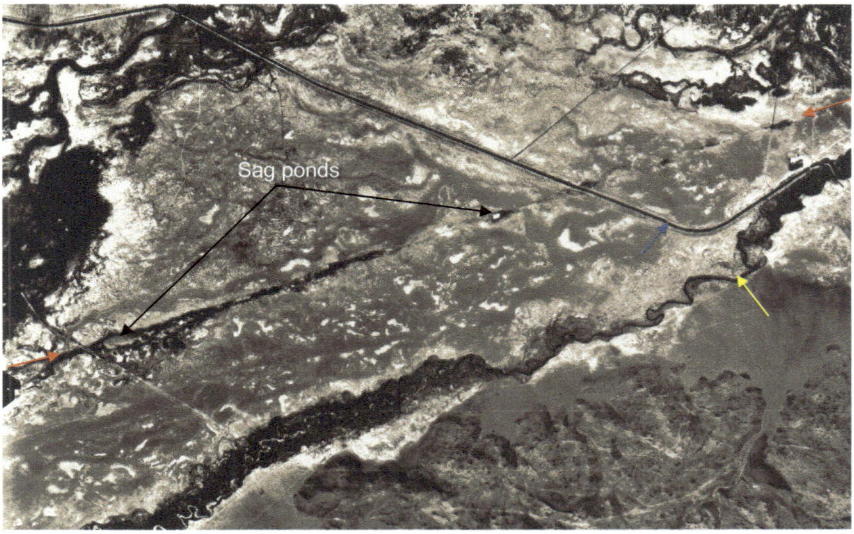

Figure 5.3 This active fault in Owens Valley, California (latest rupture in 1872) is delineated by vegetation and moisture contrasts along the trace of the fault (red arrows). The feature delineated by a blue arrow is the Los Angeles Water and Power aqueduct, which carries water from the Owens River (yellow arrow) to Los Angeles (out of the picture).

- Differences in general reflectance of the ground surface.
- Differences in vegetation distribution patterns.

5.1.3 Contrasts in Moisture

Active faults often form barriers to the free flow of water through near-surface materials or provide a conduit for flow. This variation in moisture across, or along, a fault may be large enough to be sensed remotely. The most striking effect of moisture accumulation is increased plant growth causing changes in the type, pattern, and/or density of vegetation. In areas having sparse vegetation, a moisture discontinuity may also result in changes in ground temperature due to evaporation. Moisture may change the reflectance of soils along or adjacent to the fault (refer to Figure 4.10 and Figure 5.3). Ponds may also occur along the trace of a fault forming a string of "*sag*" ponds, a characteristic feature of active faults. Note the sag ponds along the trace of the fault in Figure 5.3. Sag ponds often occur with strike-slip faults and are a classic recognition key for fault activity. Refer back to the "*location*" recognition key (number 7, Chapter 1). Why would one expect to find a string of ponds along a hillside or uphill from a major stream as seen in Figure 5.3?

Conclusions—The parameters most likely to be useful in discontinuity in ground moisture are:

- Differences in surface reflectance.
- Differences in temperature.
- Differences in vegetation type, density, or pattern.
- Occurrence of sag ponds.

5.1.4 Topographic Variations

The most definitive indicator of active faults is over-steepened land surfaces (fault scarps) as shown in Figure I.1 and the example stereo pair (page 21, Chapter 2). Fault scarps may often be detected in materials of uniform mineralogy, surface texture, or moisture conditions simply because one side of the fault, or portions within the fault zone, are closer to the camera. Since the ability to distinguish these topographic discontinuities is a powerful aid in fault detection, stereoscopic parallax is one of the most important factors to be considered when selecting a remote sensing system.

Conclusions—The following parameters are especially important for detecting topographic variations:

- Stereoscopic parallax
- Low irradiation angle
- High spatial resolution
- Thermal differences caused by shadows and highlights.

5.2 LOW SUN-ANGLE ENHANCEMENT OF SCARPS

The single most effective method for detecting and delineating fault scarps is to conduct aerial reconnaissance and remote sensing using low solar irradiation angles to produce shadows or highlights on scarps having slopes steeper than the irradiation angle of the sun. Many special studies for major engineering structures have been conducted during the past 20 years and have shown the importance of this method in revealing detailed fracture patterns associated with active faults that are difficult or impossible to observe in the field (Figure 5.4, provides a good example). Scarps having different strikes and dips can be

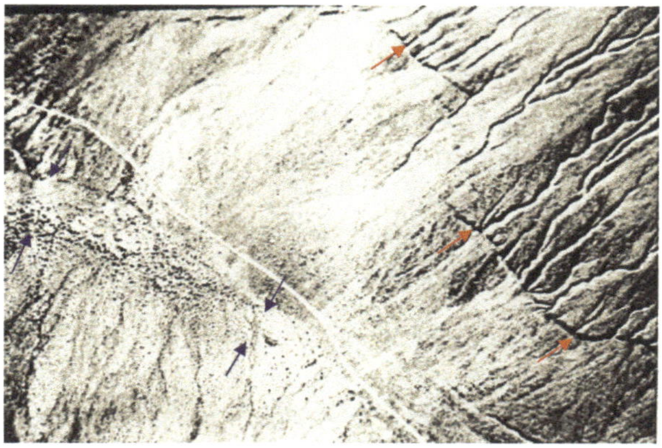

Figure 5.4 The fault scarps in Dixie Valley, Nevada (red arrows) are enhanced by low sun-angle illumination. Note the graben (blue arrows), which is easily missed even when walking along the road.

selectively enhanced using low sun-angle irradiation by photographing at the ideal time of day and year.

By planning photographic missions of specific faults according to the azimuth and scarp slope angle, it is possible to obtain images that are far superior to conventional images. I have done this for many specific faults and for areas or regions of suspected faults. I prefer to accompany the aerial photography crew in the aircraft to verify the ideal time to conduct photography.

Scarps across relatively flat terrain (floodplains, undissected alluvial fans, river terraces, plateau surfaces, etc.) should be imaged under low irradiation angles to enhance subtle features. Generally, sun angles between 10° and 25° are optimum for this type of terrain. Below a sun angle of 10° the shadows are too long for obtaining adequate detail and the light is too dim for obtaining sharp photographs. Beyond a sun angle of 25° the sun's rays are grazing most young fault scarps or irradiating older scarps. Scarps in hilly terrain (dissected plateaus, foothills, and dissected pediments) require higher illumination angles, perhaps between 20° and 35°, and in mountainous or forested regions, sun angles greater than 35° may be needed. Solar position curves, available from the US Geological Survey (USGS), portray the sun angle and azimuth for various times of the day and year for selected latitudes. Curves such as these are important for planning and reconnaissance stages of remote sensing investigations. The curves are also useful in determining approximate slope angles.

5.3 FAULT SCARP ENHANCEMENT BY STREAM EROSION

Faults that result in vertical offsets (even strike-slip faults will bring differing topography into juxtaposition, see, for example, Figure 5.13) promote accelerated stream erosion of the up-thrown block. This erosion results in flattening of the scarp crest and incision of stream channels on the up-thrown block, which stop abruptly at the scarp. Figure 5.4, shows examples of this at the red arrows.

Stream erosion along scarps and disruption of stream flow by faulting also provide clues to the presence of faults, although stream patterns are strongly influenced by other geologic structure, such as outcropping bedding planes that are moderately to steeply dipping, as well. The mere presence of a linear stream channel should not be interpreted as a fault without additional information.

A general summary of important parameters for detection of active faults on imagery should include the following:

- Differences in surface reflectance
- Stereoscopic parallax
- Low irradiation angles
- Spatial resolution
- Thermal differences
- Vegetation patterns
- Moisture differences.

Once we have identified faults on our images, it is time to determine what kind of fault we have. Why is this important? If we see it we can avoid it, right? Not so fast. Ground failure and block tilting in the vicinity of active faults can cause severe damage to structures directly, and through their exacerbating effects during strong ground shaking. For example, a structural engineer can design a building to withstand strong ground shaking as long as he knows what levels of shaking to expect. If, however, the building is tilted out of plumb and subjected to the same level of strong shaking, all design bets are off. There are two basic kinds of faults, *dip-slip faults* and *strike-slip faults*. Combinations of the two also often occur. These are termed *oblique-slip* faults. Secondary ground failure and tilting are different for each.

5.4 DIP-SLIP FAULTS

Fault dislocation along dip-slip faults occurs parallel to the dip of the fault. Two kinds of dip-slip faults are recognized, normal faults and reverse faults. The terms for these faults derives from early mining activity as shown in Figure 5.5. Mines are designed so that most of the fractures, faults, veins, and alteration zones hosting ore occur above the mine drifts. They named this area the hanging wall of the main fault, because it is unstable, dangerous, and always hanging above the heads of miners like the sword of Damocles. The footwall is the more intact and stable wall of the main fault below their feet.

There are two kinds of dip-slip faults depending upon which direction the footwall moves relative to the hanging wall. If the footwall moves upward relative to the hanging wall, the fault is known as a *normal fault*, as this is the normal case in ore deposits. If the footwall moves downward relative to the hanging wall, the fault is known as a *reverse fault* or *thrust fault*. The important thing to remember is that when a fault ruptures, the worst ground failure, fracturing, fissuring, and tilting will occur on the *hanging wall* of the fault as this is the most disturbed and unstable wall. Even if a fault is not active, you can expect the poorest foundation conditions to occur on the hanging wall.

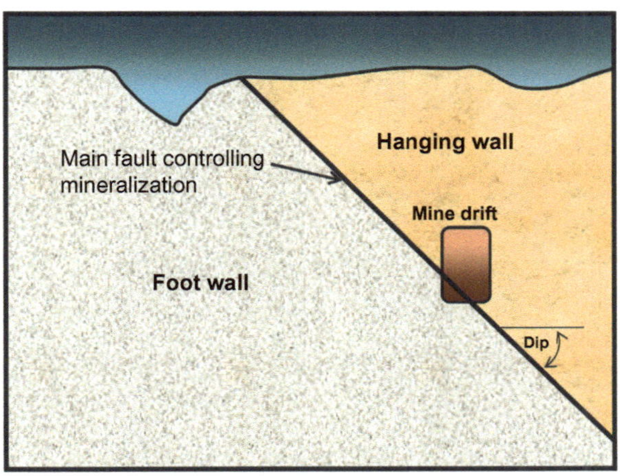

Figure 5.5 The nomenclature for dip-slip faults was established by early miners who encountered faults underground routinely as an integral part of ore deposit genesis. Fault strike is into the page.

5.4.1 Recognizing Normal Faults

Normal faults display a combination of vertical *and* extensional displacements due to the dip of the fault. Because of the extensional component, normal faults have complex fracture patterns consisting of wide zones of scarps, fissures, graben, and block tilting of the hanging wall, as shown in Figures 5.6 and 5.7.

The dip of a normal fault plane is generally steep (approximately 60°) yielding an irregular surface expression over hilly terrain. Major active normal faults probably have steep faceted fault blocks, which may show several periods of activity and quiescence. Other major landforms apparent on images include features listed in Table 5.1.

Repeated displacements along faults occasionally terminate or alter other geologic structures. These truncations, although not as obvious as some of those in Table 5.1, provide evidence of faulting, and some indication of relative age. The truncation of the glacial moraine shown in Figure 5.7A and B, for example, places the date of the last rupture of the Wasatch Fault within the Holocene (younger than 10,000 years).

Indeed, once a normal fault has been recognized on aerial photographs and the dangers associated with ground instability and tilting of the hanging wall appreciated, it is time to assess the likelihood of future characteristic earthquakes along the fault and to estimate their possible magnitudes. Detailed studies of many active faults in the

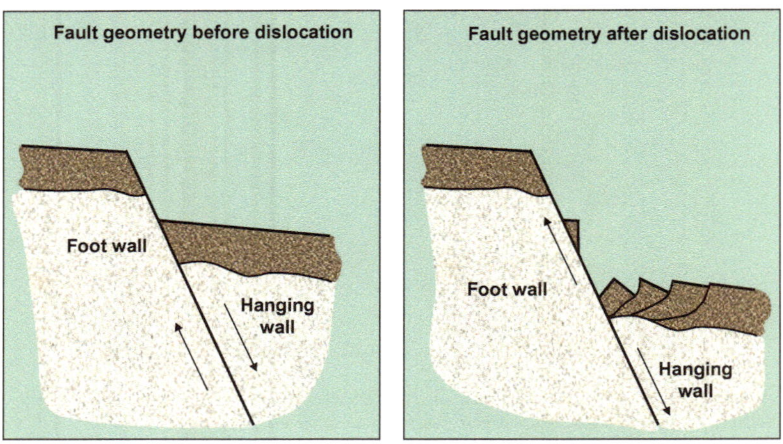

Figure 5.6 When a dip-slip fault ruptures the worst damage always occurs on the hanging wall. In the case of a normal fault, this is the down-thrown wall of the main fault.

western United States demonstrate that it is possible using evaluations of images coupled with geological, geophysical, and geochronological data to approximate the size, activity, and recurrence interval (time between fault events) to be expected along many fault zones. The imagery analysis generally is used as a first step to guide more detailed ground studies.

Offset or truncation of datable geologic features is discussed below with respect to the Wasatch Fault in Utah (Figure 5.7A and B). We may see offsets of geologic features that are larger than one would expect in a single fault rupture. In these cases, the total offset can be divided by the most likely single-event displacement indicated by field relationships, or by assumption, to yield the approximate number of events that have occurred since the formation of the geologic unit. Estimating the most likely or average single-event rupture is generally determined after a ground verification program of evaluating geomorphic features and exploratory trenching.

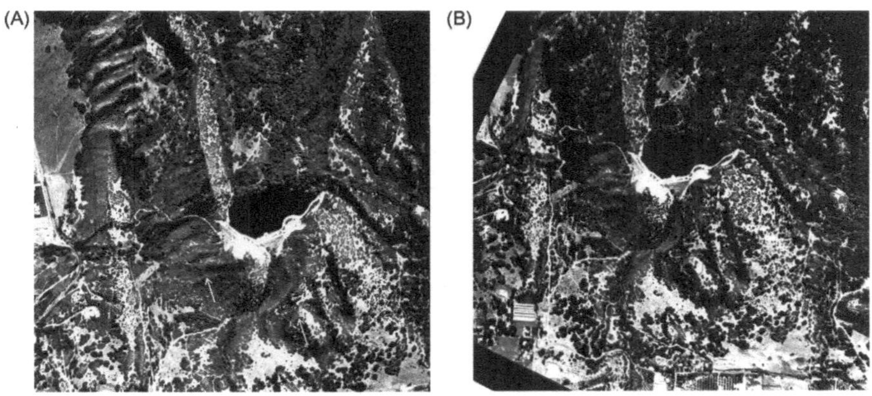

Figure 5.7 (A) This aerial photograph forms the left photograph of a stereo pair. The subject is Bell's Canyon, Utah, located approximately 15 miles south of Salt Lake City. Little Cottonwood Canyon is at the left edge of the photograph. The main Wasatch Fault scarp (in shadow forming the east side of the small reservoir) is a normal fault that ruptured prehistorically. The curved feature forming the boundary of most of the rest of the reservoir is a glacial moraine. Note the complexity of the faulting on the down-thrown side (hanging wall to the west or bottom of the photograph) of the fault. Back facing scarps are highlighted by the sun. I have placed a few arrows to delimit some of them and to provide some idea of the width of the zone of damage. (B) This aerial photograph forms the right photograph of a stereo pair. The subject is Bell's Canyon, Utah, located approximately 15 miles south of Salt Lake City. Little Cottonwood Canyon is at the left edge of the photograph. You may have to adjust the photograph under the stereoscope a little as I had asked the pilot to turn the airplane to follow the trace of the Wasatch Fault. I have attempted to correct for the turn, but may not have adjusted it perfectly.

Table 5.1 Value of Different Image Scales[a] for Recognizing Landforms of Normal Faults

Geomorphic Feature	Synoptic Scale/Topographic Maps	Intermediate Scale	Large Scale
Scarps			
Simple scarps	Yes for large scarps	Yes	Yes
Fissures	No	Probably not	Yes
Graben or trenches	Large scarps, sparse vegetation, low sun	Probably	Yes
En echelon scarps	Large scarps, sparse vegetation, low sun	Probably	Yes
Subsidence fissures	No	No	Yes
Beveled scarps	No	No	Yes
Triangular facets and faceted spurs	Yes	Yes	Yes
Orthogonal fracture systems	Yes/Doubtful	Yes	Yes
Arcuate scarps (at terminations or in sediment)	No	No	Yes
Rejuvenated terraces	No	Yes	Yes
Wing glass canyons	Yes	Yes	Yes
Alluvial aprons, fans	Perhaps/Probably	Probably	Yes
Ramps	No	Doubtful	Yes
Groundwater effects (linear springs, differential water tables)	Probably/Perhaps	Yes	Yes
Volcanic features	Yes	Yes	Yes
Mountain front embayment	Yes	Yes	Yes
Canyon ellipticity	Yes	Yes	Yes
Valley ratios	Doubtful/Yes	Yes	Yes

[a]Synoptic scale ≤ 1:75,000; 1:25,000 ≥ Intermediate scale ≥ :75,000; Large scale ≥ 1:25,000.

5.4.1.1 Erosion of Fault Scarps

Erosion differences across and along scarps may provide an additional means of estimating ages of faulting events and the typical rupture offsets per event. Principal characteristics of young scarps include steep free face, debris slope, and a sharp break in slope at the crest of the scarp. Scarps associated with older displacements tend to exhibit lower slopes resulting in broader expression on images. Along faults that have had repeated displacements, scarps exhibit a composite or multiple-slope form such as that shown in Figure 5.8.

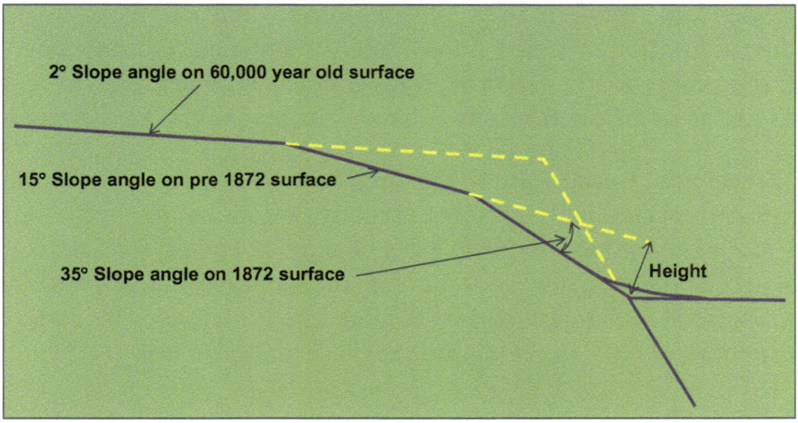

Figure 5.8 The diagram shows a beveled fault scarp for the 1872 earthquake in Owens Valley, California. A minimum of two large previous events occurred along the same fault scarp.

Three ways to use simple measurements of the fault scarp geometry to estimate the age of the different periods of activity are *regression analysis*, *diffusion modeling*, and *canonical analysis*.

In regression analysis one measures the slope angle and height of the fault scarp at several locations along its strike. These measurements are then plotted as Max[slope] versus Log[Height] and compared with similar plots for scarps having known ages.

Diffusion modeling is an empirical approach to quantifying fault scarp degradation proposed by Wallace[2] and refined by many other authors for other regions since. The strategy uses a diffusion equation given by

$$T_{\mathrm{u}} = T_{\mathrm{d}}\left(\frac{H_{\mathrm{u}}}{H_{\mathrm{d}}}\right)^2$$

where T_{u} is the unknown age of the scarp being estimated, T_{d} is the age of a known scarp, and H is the scarp height as shown in Figure 5.8.

A canonical analysis is essentially a principal components approach to maximize the discrimination of young scarps in some feature space (the Max[slope] versus Log[Height] space works just fine). Think of a 2-D plot of Max[slope] versus Log[Height] with a selection of points for a fault having known age along with others having unknown ages within the same climatic area. The canonical analysis seeks a simple coordinate rotation to a new coordinate system such that one axis falls along the points for the fault of known age. Older scarps will be displayed above or below this axis a distance proportional to their ages.

5.4.1.2 Tectonic Adjustments to Fluvial Processes

Records of repeated fault activity over long periods of time indicate that many faults do not have uniform rates of activity; rather, they show periods of accelerated activity followed by periods of quiescence. These intermittent periods of activity may be distinguished on images by variations in the amount of erosion that has taken place along a scarp.

Geomorphic evidence of late Pleistocene and Holocene movements along faults includes fault ruptures of: (a) alluvial fans, (b) strath terraces, (c) playas, and (d) rock falls from fault scarps onto Pleistocene or Holocene surfaces. In areas where scarps may not be present, Bull and McFadden[3] have developed a fault classification system based on the adjustments that tectonic activity causes to fluvial systems crossing the fault. Three quantitative metrics together with numerous qualitative observations are used to classify mountain front tectonic activity. The three quantitative metrics are (a) mountain front sinuosity, (b) valley width to height ratio, and (c) drainage basin ellipticity.

Mountain front sinuosity: Streams flowing across a tectonically active mountain front erode downward as the mountain block rises. They have little time to establish lateral erosion or meandering before the next tectonic uplift. As tectonic activity wanes along a mountain front, however, streams flowing across the front begin to erode laterally, develop meanders, and form floodplains. This erosional and depositional behavior causes embayments to form into the mountain block. These embayments cause the mountain front boundary to become more sinuous. As tectonic quiescence progresses, so does the sinuosity.

To compute the sinuosity of a mountain front, one assumes a smooth curve for the surface trace of the range-bounding fault. The length of this curve is L_f, the approximate length of the range-bounding fault. Next, one traces the irregular bedrock-alluvium contact. The length of this trace is L_{mf}, the length of the mountain front. This length is obviously a fractal, so it will increase as the measurements become more precise. Choose an appropriate measurement precision for the image or map scales in your study, and stick with it. Sinuosity is then defined as

$$S = \frac{L_{\mathrm{mf}}}{L_{\mathrm{f}}}$$

Valley width to valley height ratio: This metric takes advantage of the observation made above that streams crossing tectonically active mountain fronts have little time for lateral erosion, whereas streams crossing less active tectonic fronts meander and develop floodplains. To compute this ratio, choose a few of the major streams exiting the mountain front. Measure the average height of the valley sides near the mountain front (within about 1 km usually, but be consistent). This measurement is V_{h}. At the same location, measure the width of the valley floor from where the bedrock meets the flood plane on one side to the same intersection on the other side. This measurement is V_{fw}. The ratio is then defined as

$$V = \frac{V_{\mathrm{fw}}}{V_{\mathrm{h}}}$$

Drainage basin ellipticity: This measurement embraces the entire drainage basin for major streams within the mountain block. As tectonic activity wanes the drainage basin shape widens from a sharp ellipse to a more circular shape. To compute this ratio, choose the same stream valleys you used to compute the valley width to valley height ratio. Measure the area, A_{d}, of the drainage basin. If the drainage basin were circular, its diameter, D_{d} would be approximately equal to

$$D_{\mathrm{d}} \approx 1.13\sqrt{A_{\mathrm{d}}}$$

Now measure the distance L_{d} between the two most distant points in the drainage basin, and compute the ellipticity as

$$E_{\mathrm{db}} = \frac{D_{\mathrm{d}}}{L_{\mathrm{d}}} = \frac{1.13\sqrt{A_{\mathrm{d}}}}{L_{\mathrm{d}}}$$

Although the metrics will change depending on the scale and precision of the images or maps, the following guidelines were presented by Bull and McFadden:

Class I (tectonically active): These terrains generally have a mountain front sinuosity ranging from 1.0 to 1.6, unentrenched alluvial

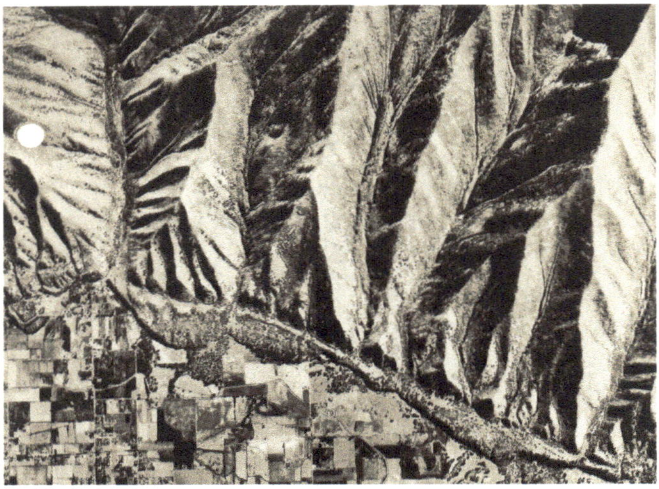

Figure 5.9 The Class I mountain front is located along the Wasatch Fault south of Provo Utah. Note the triangular facets along the mountain front. The differing sizes of facets indicate periods of tectonic activity followed by periods of tectonic quiescence. The larger streams have steep valleys and each occupies the entire valley floor. Note the elliptical drainage basins and the lack of embayment at the mountain front. Sinuosity is nearly 1.0.

fans with the newest sediment deposits forming steep cones at the mountain front boundary with the apex of the alluvial fans, elongated drainage basins having narrow valley floors and steep hill slopes. Class I terrains usually have fault scarps offsetting Pleistocene or Holocene surfaces (Figure 5.9).

Class II (moderate to slightly active): These terrains have a sinuosity ranging from 1.4 to 3.0, permanent entrenchment of alluvial fans with the newest sediment deposits forming farther down the fans, large drainage basins that are more circular than Class I , steep hill slopes, and valley floors that are wider than the flood plain. Class II terrains generally show activity during the Pleistocene, but not the Holocene (Figure 5.10).

Class III (tectonically inactive): These terrains generally have a sinuosity ranging from 2.0 to greater than 7, pedimented mountain fronts and embayments, and few large integrated stream channels in mountains. Steep slopes are associated only with resistant rocks (Figure 5.11).

The beauty of this approach is that small scale images (Landsat and Google Earth, for example) or topographic maps are amenable to the task. Comparisons of the tectonic activity of mountain fronts can be made over regional scales as long as climate conditions are

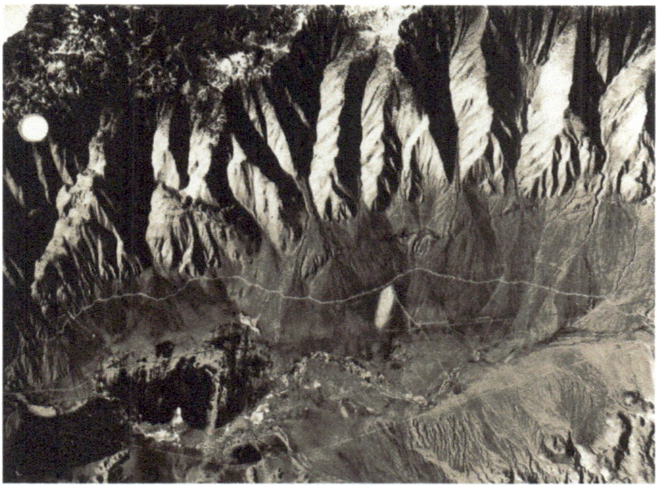

Figure 5.10 The Class II tectonic front is located along the eastern edge of the Sierra Nevada Mountains in southern Owens Valley, California. The mountain front shows embayment and deeply incised alluvial fans with the youngest alluvium being deposited near the base of the fans. Yet there are young scarps offsetting the fans indicating renewal of tectonic activity.

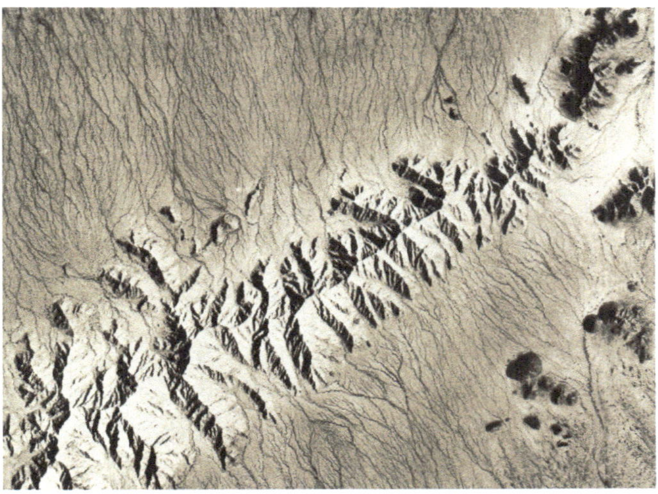

Figure 5.11 The Class III mountain front is deeply embayed, sediment is carried well beyond the mountain front and deposited in streams in the middle of the valley, and steep slopes are confined to the core of the range. Faulting toward the bottom of the image probably postdates that toward the top.

similar over the area of study. I have used this technique, for example, as an aid to establishing the least hazardous route for pipelines from Wyoming to California across the numerous faults of the

Great Basin[4] and to assess fault activity in New Guinea using Google Earth images.

5.4.1.3 Fault Segmentation

Fault segmentation is related to the size of asperities likely to rupture during a characteristic earthquake. This is an important feature to understand if we are to estimate the size of potential future earthquakes. It is rare for an entire fault to rupture during a single characteristic earthquake. Factors that combine to determine asperity size and strength, and hence fault segmentation include (a) geometric irregularities, (b) stress concentrations, and (c) fluid pressure. Aerial photography can be used for determining fault segmentation in the following ways: (a) comparing mountain front characteristics as discussed earlier and (b) analysis of fault scarp freshness and continuity as discussed earlier. Large fault ruptures are certainly capable of sweeping through several asperities, but only rarely. It's more common for a single segment or asperity to rupture and create a characteristic earthquake, so pay close attention to the length of segments within which scarp characteristics are similar, and note the scarp height (or displacement in the case of a strike-slip or oblique-slip offset) as this may be the only indicator you will have of potential fault displacement. Why is this important? It is important because it provides you with a way to estimate the size of earthquake that a fault segment can produce. An estimate of earthquake size can be made using the *Seismic Moment*. The seismic moment is given by

$$M_0 = \mu A \overline{u}$$

In the above equation, M_0 is the seismic moment in dyne-cm, μ is the shear strength of the asperity in dyne-cm^{-2}, A is the area of the asperity in cm^2, and $\overline{u}$ is the amount of fault displacement in cm. Reasonable estimates for these values are:

1. Crustal shear strength, μ, is on the order of 3×10^{11} dyne-cm^{-2}. Use this for the shear strength of the asperity if you have no other information.
2. Asperity area A is the asperity length L times the asperity width (depth), W. A good value for the asperity width, W, is approximately 17 km for the western United States, and the asperity length, L, is the segment length that you have just measured. Of course you

need the area in centimeter square and the length in centimeter to get the proper units for the seismic moment.

3. The fault displacement, $\overline{u}$, is the scarp displacement in centimeter that you have measured.

Once the seismic moment has been calculated you can convert this to an estimate of potential moment magnitude, M, using the equation

$$M = \frac{2}{3}\mathrm{Log}_{10}M_0 - 10.7$$

The discussion of fault segmentation, rupture size, seismic moment, and moment magnitude are consistent with any type of fault whether they be dip slip, strike slip, or oblique slip.

5.4.2 Recognizing Reverse Faults

Reverse fault displacements combine vertical and compressional displacements. Reverse faults typically have a wide range of dip angles. Reverse faults with low dips exhibit a sinuous surface expression over hilly to flat terrain. Reverse fault scarps are often difficult to locate precisely due to widespread landslides which cover the fault trace. Surface fracturing is characterized by folding or distortion and displacements along subsidiary faults and fractures within the hanging wall (Figure 5.12A).

Fracturing and ground instability on the hanging wall of a reverse fault is commonly spread over a wide area (tens of miles in some cases), more than is common with normal-slip and strike-slip faults due mostly to the amount of frictional forces involved, which in turn partition or disperse stresses over greater distances depending generally on rates of movement, the rock types involved, and the presence or absence of fluids.

Figure 5.12B shows a thrust fault rupture resulting in the 1968 magnitude 6.9 Meckering WA, Australia (approximately 130 km east of Perth) earthquake. Note on Figure 5.12B the wide area of damage due to secondary scarps and slumps on the hanging wall of the fault (left on the photograph). The earthquake caused ground rupture of nearly 40 km, with a 2.4 m vertical offset and a 1.5 m horizontal offset. Although the town of Meckering was destroyed during the earthquake, none of its citizens was killed.

The main geomorphic features of reverse-slip faults are listed in Table 5.2.

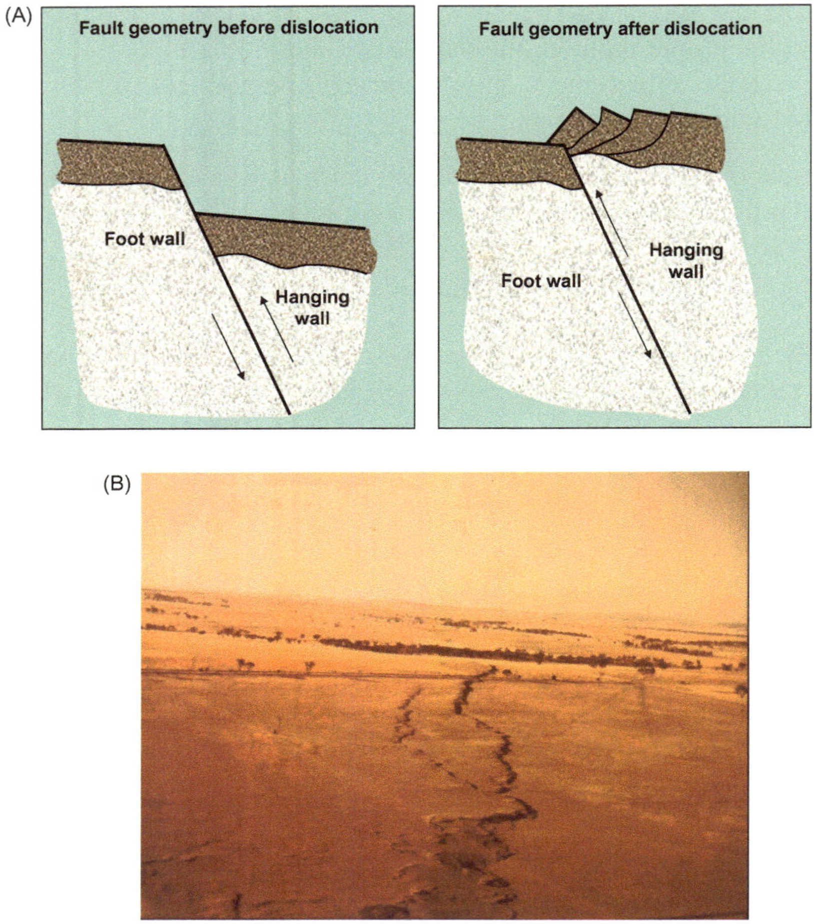

Figure 5.12 (A) Reverse faults display severe damage in the form of landslides over the fault trace caused by the inability of the hanging wall to support the overhang caused by the fault displacement, folds, and compression features within the fractured hanging wall, and compressional block tilting. (B) Thrust fault scarp due to rupture causing the 1968 magnitude 6.9 Meckering WA, Australia earthquake. Photograph courtesy of Ian Everingham, Peter Gregson, the West Australian newspaper, Alice Snocke, and Wayne and Brenden Peck.

5.4.3 Recognizing Strike-Slip Faults

Strike-slip faults exhibit horizontal movement along the strike of the fault with little vertical movement along the dip. Strike-slip fault planes generally have a nearly vertical dip and exhibit linear or gently curved fault traces on flat and hilly terrain. Fault breaks tend to be relatively simple with some branching or secondary scarps along the principal displacement zone, although there is usually a wide, complex

Table 5.2 Value of Different Image Scales[a] for Recognizing Landforms of Reverse Faults

Geomorphic Feature	Synoptic Scale/Topographic Maps	Intermediate Scale	Large Scale
Scarps	Yes for large scarps	Yes	Yes
Blunting or oversteepening at foot of mountain front	Yes	Yes	Yes
Talus or landslide alignments	Possible/Doubtful	Yes	Yes
Mole-track traces	No	Probably	Yes
Graben or fissure swarms on hanging wall	No	Yes	Yes
Upstream terraces	No	Perhaps if large	Yes
Drag warping of footwall fans, terraces, or sediment	No	Doubtful	Yes
Sinuous traces on flat surfaces	Possible if large/No	Doubtful	Yes
Volcanic features	Yes	Yes	Yes
Mountain front embayment	Yes	Yes	Yes
Canyon ellipticity	Yes	Yes	Yes
Valley ratios	Doubtful/Yes	Yes	Yes

[a]*Synoptic scale* $\leq$ *1:75,000; 1:25,000* $\geq$ *Intermediate scale* $\geq$ *1:75,000; Large scale* $\geq$ *1:25,000.*

zone of faults, fractures and graben and compressional folds, particularly where the strike-slip fault varies from linear. Figure 5.13 shows some of the most common geomorphic features of strike-slip faults. En echelon and Riedel patterns are common. Typical geomorphic features are presented in Table 5.3, and an example of strike-slip faulting is shown on Figure 5.14.

Elements that favor detection of strike-slip faults on images include:

- High spatial resolution
- Stereoscopic coverage
- Low irradiation angles.

Once faults have been identified other hazards from earthquakes can be addressed, remember, there were four others.

5.5 IDENTIFICATION OF GROUND SHAKING HAZARD

I think everyone would agree that placing a structure astride an active fault is not a good idea. Depending on the type of fault, careful ground

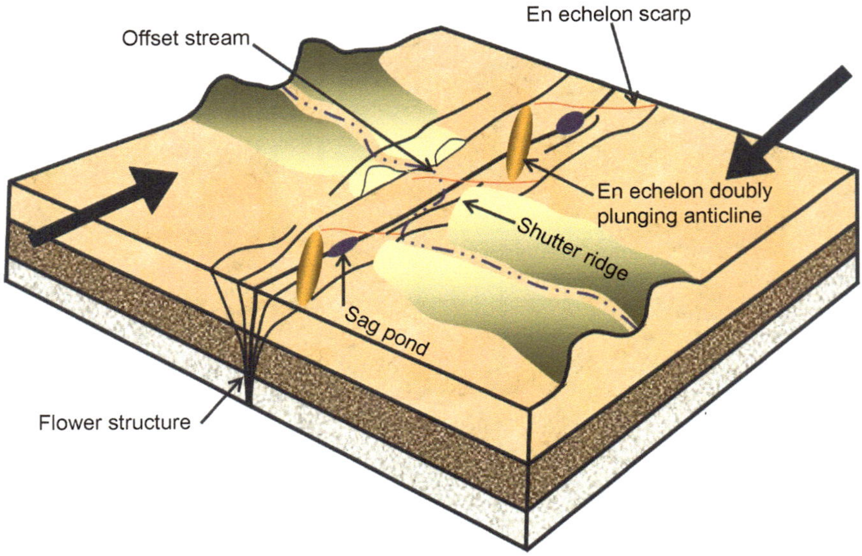

Figure 5.13 Typical geomorphic features associated with strike-slip faults include, sag ponds (see Figure 5.3, for example), complex zone of faulting as a result of the flower structure typical of strike-slip faults, offset streams, shutter ridges, en echelon scarps, and en echelon doubly plunging anticlines (brachyanticlines). Note the different directions for the major axes of the anticlines and the en echelon fractures. The large arrows indicate a right-lateral sense of movement of the fault.

studies including mapping, trenching, and geophysics should be completed, within perhaps a broad zone adjacent to the main fault to account for ancillary ground failure due to hanging wall damage, tilting, en echelon faults, and other flower structure induced secondary faults. Once these contingencies have been successfully evaluated and avoided, mere proximity to an active fault does not necessarily indicate higher hazard. Evaluating dangers from ground shaking is largely beyond what we can accomplish using aerial photographs as it requires information on earthquake size (something we *can* approximate using fault segmentation and scarp height discussed earlier), elastic and inelastic attenuation of earthquake wave amplitudes and frequencies between the earthquake source and specific sites, and soil geometry and elastic properties at specific sites.

What we *can* delineate using aerial photographs are areas that are likely to have deep deposits of poorly consolidated sediment vis a vis well-indurated rock at, or close to, the surface. These delineations provide assistance to engineers for designing site exploration programs to

Table 5.3 Value of Different Image Scales[a] for Recognizing Landforms of Strike-Slip Fault Systems

Geomorphic Feature	Synoptic Scale/Topographic Maps	Intermediate Scale	Large Scale
Scarps	Yes	Yes	Yes
Linear canyons	Yes	Yes	Yes
Sag ponds	Possible/Doubtful	Yes	Yes
Offset streams	Yes	Yes	Yes
Fault gap or notch	If large enough	Yes	Yes
Trench or wedge	If large enough	Perhaps if large	Yes
Shutter ridges	If large enough	If large enough	Yes
Vegetation alignments	Yes/Doubtful	Yes	Yes
Volcanic features	Yes	Yes	Yes
Brachyanticlines	Yes	Yes	Yes
Side hill trough	Doubtful	Yes	Yes
Fault valley	Yes/Yes	Yes	Yes

[a]*Synoptic scale ≤ 1:75,000; 1:25,000 ≥ Intermediate scale ≥ 1:75,000; Large scale ≥ 1:25,000.*

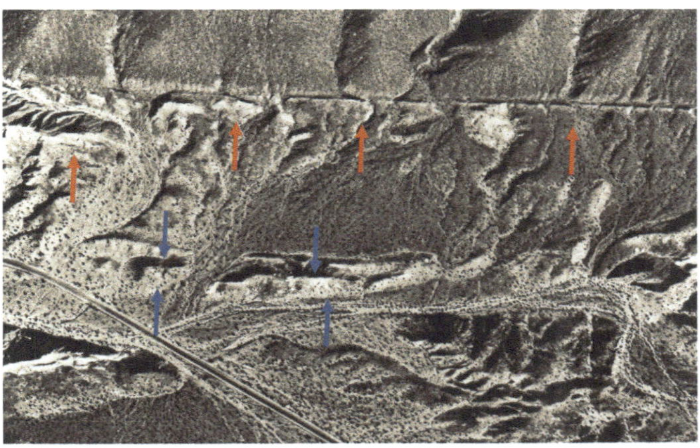

Figure 5.14 The vertical aerial photograph is of the Garlock fault, California. North is to the top of the photograph. Note the shutter ridges forcing streams to bend as they cross the principal displacement zone (red arrows). Strike-slip faults usually have steep dips and, therefore, may appear ruler straight over even rough terrain. You can see other, older, scarps near the bottom of the photograph (blue arrows) due to the wide zone of faulting related to the flower structure common to strike-slip faults.

assess the soil geometry and elastic properties necessary to evaluate ground shaking hazard.

5.6 IDENTIFICATION OF LANDSLIDE HAZARD

Landslides are common during earthquakes, and the damage caused by them can, in some instances, exceed damage due to earthquake ground shaking and fault displacement[5]. The Northridge, California earthquake in 1994, for example, triggered more than 11,000 landslides[6].

Assessment of landslide hazard requires knowledge of landslide geometry, soil or rock strength and frictional properties, loading conditions (both static and dynamic), and pore water pressure distribution within the slide mass and antecedent rainfall. Few of these are available from aerial photographs; most require detailed surface and subsurface interrogation, and even then may not be realizable. Some landslides are simply unpredictable, defying our most sincere analyses.

We certainly *can*, however, recognize landslides that have already occurred. If we see a soil or rock unit that shows evidence of past landslides, that should be to us like a matador waving a red cape is to a bull. It tells us that here is a unit that is potentially dangerous to build upon. We can then notify site engineers that special care is needed. Since recognizing landslides is the subject of Chapter 6, we shall consider that subject in more detail there.

5.7 IDENTIFICATION OF LIQUEFACTION HAZARD

Liquefaction occurs in loose, saturated granular (sand mostly) materials subjected to prolonged strong ground shaking. As the loose sand attempts to compact in response to earthquake vibrations, water in the pores can't escape, so pore pressures begin to increase. As vibration continues, pore pressure continues to increase until the upward pressure of the water equals or exceeds the downward forces of the soil, resulting in sandy water rather than watery sand. Sandy water can not sustain shear forces, hence any structures built on the liquefying sand sustain damage.

The operative conditions are strong, prolonged ground shaking, and saturated, granular soil. As aerial photo analysts, we are concerned here mostly with saturated, granular soil in the vicinity of our largest characteristic earthquakes. Small earthquakes simply do not have the wave amplitudes or duration of strong ground shaking necessary to produce the increased pore pressures needed for liquefaction. On aerial photographs, we flag the obvious lakeshores, bay-shores, stream shores, and low areas in the center of valleys, all areas likely to have saturated, loosely compacted granular soil. Evidence of past liquefaction includes broad areas of ground failure, sand volcanoes, sand boils, and sand-delineated fissures. The New Madrid earthquake of 1833 created sand boils and fissures easily identified on aerial photographs of farm land well away from the Mississippi and Missouri Rivers in Missouri and Arkansas. Low areas may also consist of thick bogs having saturated interbedded organic clay with silt or sand lenses. During strong earthquake shaking the sand lenses can liquefy with a resulting decrease in overall ground stability. If the ground is sloping, large down-slope ground translations can occur similar to the damaging ground failure at Turnagain Heights, Alaska during the 1964 earthquake.

It is useful to remember that, even though shaking is the best way to compact loose, dry sand, shaking is *not effective* in compacting loose, saturated sand. Liquefied sediment remains a liquefaction danger until it is dewatered and compacted properly.

5.8 IDENTIFICATION OF TSUNAMI AND SEICHE HAZARD

Tsunami are seismically generated sea waves. Most coastal areas are subjected to tsunami hazard. Tsunami generation requires a large vertical offset of the seafloor. Once generated, tsunami propagate through deep water at high rates of speed, long wavelengths, and low amplitudes to distant shores. Proximity to the earthquake source is not a useful criterion for tsunami hazard. A tsunami generated by a magnitude 8.2 earthquake off of the Kamchatka Peninsula, USSR, Figure 5.15, caused approximately $1M in property damage to the Hawaiian Islands,

Coastal areas facing, or in line with, active subduction earthquake zones are especially susceptible to tsunami damage. This places circum-Pacific coastal areas, islands within the Pacific and Indian Oceans, and coastal areas around the Indian Ocean at risk for tsunami. Even though the Atlantic Ocean and other oceans and seas have

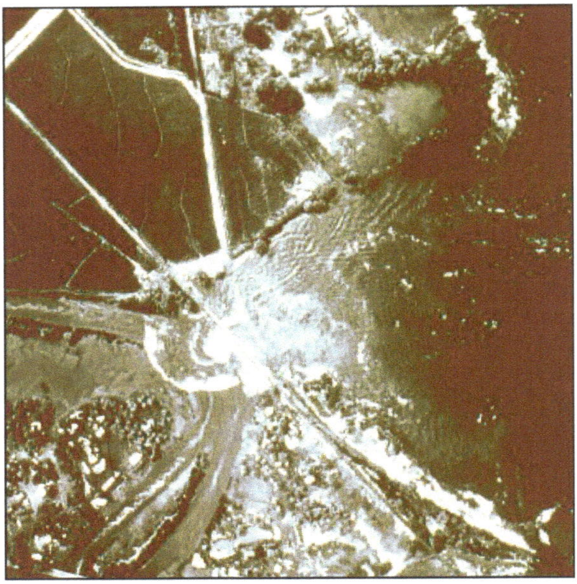

Figure 5.15 On November 4, 1952 this tsunami damaged Kaika Bay near Haleiva, on the north shore of Oahu due to a magnitude 8.2 earthquake off the Kamchatka Peninsula, USSR. Photo courtesy of the National Geophysical Data Center.

not generated large tsunami historically, an earthquake off the coast of Hawaii on December 29, 1975 did generate a small tsunami, and large volcanic explosions (the 1883 eruption of Krakatau, for example, generated a tsunami which inundated the coast of Java killing 36,000 people) are also known to have caused devastating tsunami. In addition, large waves have been reported in Italy from landslides into the Mediterranean. This should be considered wherever steep cliffs indicate the possibility of large landslides into water bodies (offshore sonar around the Hawaiian Islands, for example, indicates large, thick accumulations of landslide debris, some of which could have generated large waves on nearby islands).

Tsunami run-up is a function of coastal and offshore geometry as well as the orientation of a site with respect to potential tsunamigenic zones, the geometry of tsunamigenic fault zones, and the size of seabed ruptures. Many pictures and film of the 2011 Tōhoku, Japan magnitude 9.03 earthquake and accompanying tsunami provide insight into the nature and dangers to coastal areas from tsunami.

Seiches are generated in fully or partially enclosed bodies of water such as lakes and bays when the frequency content of incoming seismic waves matches the resonant characteristics of the water body. Visualize the back and forth sloshing that can occur by shaking a tub of water. A seiche in Hebgen Lake during the 1959 Hebgen Lake earthquake near West Yellowstone, Idaho overtopped the Hebgen Lake dam several times. This seiche was compounded by strong ground shaking from the earthquake, the lake's location on the hanging wall of the ruptured normal fault, which caused dropping and tilting of the lake floor, and a small landslide into the lake. All lakes and bays within the vicinity of potentially large characteristic earthquakes should be flagged as seiche hazards.

REFERENCES

1. Casavant RR. 2010, Personal Communication.
2. Wallace R. Profiles and ages of young fault scarps, north central Nevada. Geol Soc Am Bull 1977;88:1267–81.
3. Bull WD, McFadden LD. Tectonic geomorphology north and south of the Garlock Fault, California. In: Doering DO, editor. Geomorphology in Arid Regions, A Proceedings Volume of the Eighth Annual Geomorphology Symposium held at SUNY Binghamton; 1977, 23–24.
4. Tula A, Glass CE. Regional active fault assessment using tectonic geomorphology—an example from the rocky mountain pipeline, 76th Annual Meeting of the Cordilleran Section of GSA, Abstracts with Programs; 1980, p. 156, 19–21.
5. Jibson RW, Harp EL, Michael JA. A method for producing digital probabilistic seismic landslide hazard maps and example from the Los Angeles, California area USGS Open File Report 98–113; 1998, 22p.
6. U.S.G.S. USGS response to an urban earthquake—Northridge '94. USGS Open File Report 96–263; 1996, p. 43–47.

CHAPTER 6

Dangers from Landslides

Landslides cause \$1–2 billion damage and up to 50 deaths annually in the United States. That's more than all other natural hazards combined. No one is immune to them. They routinely destroy single family homes equally in wealthy and disadvantaged neighborhoods as well as commercial real estate, and damage infrastructure (Figure 6.1).

One of the most important indications of potential landslides is the existence of areas having oversteepened slopes, formations known to comprise unstable material, and geologic units having geomorphic expressions of previous landslides. Thus, recognizing an old landslide when you see one gets you a long way toward avoiding, or if possible, stabilizing landslide-prone areas. As Figure 6.1 unquestionably demonstrates, you *can* see landslides on aerial photographs. So why in the world would you not use them? I can guarantee you this, if you build on a landslide without knowing it, and future sliding damages structures, you will find yourself in court sitting across from someone like me who has an enlarged aerial photograph showing to the jury the landslides you missed.

Assessment of landslide hazard requires knowledge of landslide geometry, soil, or rock strength and frictional properties, loading conditions (both static and dynamic), antecedent rainfall, and pore water pressure distribution within the slide mass. Few of these are available from aerial photographs; most require detailed surface and subsurface interrogation, and even then may not be realizable. Some landslides are simply unpredictable, defying our most sincere analyses.

We certainly *can*, however, recognize landslides that have already occurred. If we see a soil or rock unit that shows evidence of past landslides, that should be to us like a matador waving a red cape is to a bull. It tells us that here is a unit that is potentially dangerous to build on. We can then notify site engineers that special care and design will be needed. So what do landslides look like? That depends on whether they occur in soil or rock. For this discussion, the term "*soil*," comprises all materials in which the slope failure occurs primarily through intact soil or soft rock media, and only secondarily along joints or faults. This includes residual soil, deposited sand, silt and clay, poorly indurated shale, and friable sandstone. Let's look at landslides in soil first, then later cover landslides in rock in which the slope failure occurs primarily along joints or faults and only secondarily through intact material.

6.1 LANDSLIDES IN SOIL

Figure 6.1 shows many of the features typical of landslides in soil. The geologic units affected by the landslides involve Pleistocene lakeshore deposits. Arcuate scarps are common and look like giant bites taken out of headlands. These can readily be seen in Figure 6.1 near the neighborhoods to the south of the landslides (yellow arrows). Slump blocks within the landslide debris in Figure 6.1 are covered with trees, but the blocks are there. It doesn't take much time doing stereoscopic analysis of aerial photographs of that site to decide that the landslides

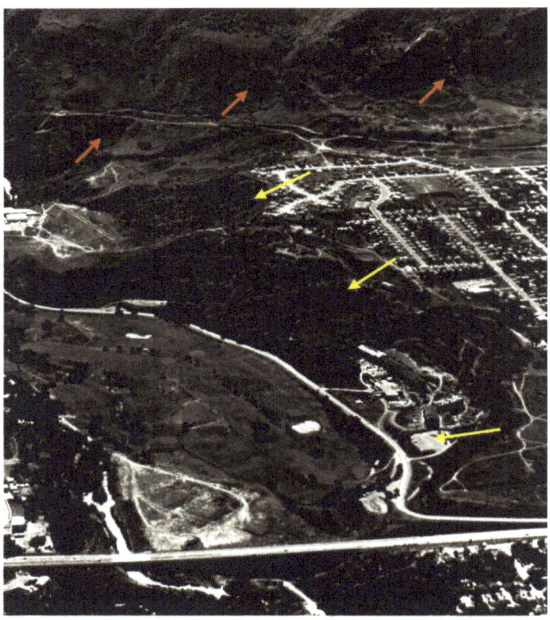

Figure 6.1 The oblique aerial photograph looks east toward the mouth of Ogden Canyon, Utah. There is a series of landslides south of the golf course (yellow arrows), with new condominium construction on the landslide debris. The golf course is also built on landslide debris. Is this acceptable land use? The mountains in the background are part of the Wasatch mountain range. After reading Chapter 5, can you identify active fault scarps associated with the Wasatch Fault (footwall shown by red arrows)? If so, why do you think the scarps suggest activity?

are active. One quickly notes, for example, that trees on the slump blocks tilt down-slope. Unless one is willing to believe these are Pleistocene era trees, movement must have occurred during the life of the tree. Also noticeable is disruption to the parking lot of the old wedding reception chapel at the westernmost yellow arrow. As if these indications aren't enough, signs along the road alert motorists to a "*Landslide Area*".

Figure 6.2 shows common features of landslides. Most landslides in shallow soil have shallow planar failure surfaces similar to the smaller slides shown in Figure 6.2. As the depth of soil increases the failure surface becomes more curved, assuming a logarithmic spiral shape similar to the large landslide in Figure 6.2, then a circular shape for the thickest soil.

Landslide debris is hummocky, irregular, and often forms ponds due to the disruption of groundwater flow by the landslide. Landslide toes are globular, frequently forming lobes. Take a look at the large landslides near Corfu, Washington in Figure 6.3.

Now note the debris flow on Figure 6.2. Observations suggest that the majority of cold debris flows (those not generated directly by volcanic eruptions) begin as shallow, slab-like landslides in saturated surface soils. The landslide gains speed rapidly as it moves downhill where it finally impacts a stream channel. This impact usually scours

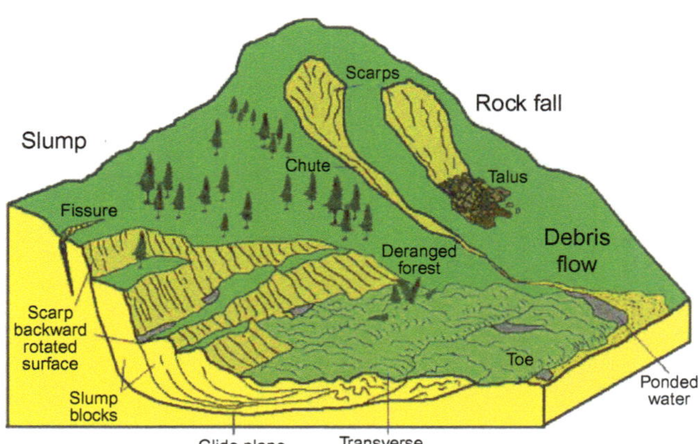

Figure 6.2 The cartoon displays soil landslide geometries. This excellent diagram has been made available to the public together with a concise description of different slide types by the USGS on their website http://3dparks.wr.usgs.gov/landslide/types.htm.

stream channel deposits containing water and, if it hasn't already, the slide becomes a mobile, viscous debris flow (a Bingham flow[1]). The new debris flow forms a deformation zone (boundary layer) at its base. The material above the deformation zone forms a semi-rigid "*plug*" in the center and top of the flow. This plug may translate downstream relatively undisturbed, which explains observations of trees, soil, and an occasional human riding the center of a large debris flow undisturbed for several kilometers. Antecedent rainfall has an important effect on the triggering of debris flows, so consider local rainfall records as part of your hazard analysis.

As debris flows are capable of causing damage and loss of life several kilometers from the source of the debris flow, it is important to be able to recognize debris flow hazard. Some of the evidence of debris flow run-out is visible on aerial photographs where extensive cultural activity has not yet destroyed it. Unfortunately, there is a tendency, which I call the "*Pioneer Syndrome*," to rush back into an area devastated by a landslide, debris flow, or flood and quickly rebuild. The pioneer spirit encourages us to not let nature get us down, even when we should. So it is important to look not only at aerial photographs taken recently, but also at photographs of the same area taken over time. Don't get tunnel vision and look only at your particular site. Take a synoptic view, even if it costs a little more to acquire extra photographs. Don't miss the landslides and active fault scarps displayed on Figure 6.1 just because your particular site is a few blocks away. As an

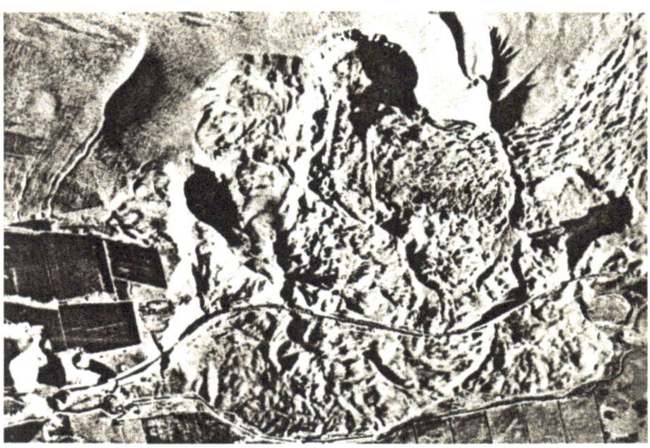

Figure 6.3 The vertical aerial photograph shows a series of landslides near Corfu, Washington. Note the characteristic circular landslide head scarps and the disruption typical of landslide debris.

example, Figure 6.4 shows an aerial oblique photograph of a debris flow caused by failure of weak, hydrothermally altered rock on Casita Volcano in Nicaragua during hurricane Mitch on November 2, 1998. Note the many other indications of ground instability.

The stereo pairs shown in Figures 6.5 and 6.6 provide some examples to look at.

6.2 LANDSLIDES IN ROCK

Landslides in rock look a little different than landslides in soil. Most landslides in rock occur along discontinuities (faults or fractures) in the rock mass, and the size of the slide depends largely on the sizes of the rock blocks between fractures. The slide plane can be planar, where failure occurs along a planar discontinuity (top of Figure 6.7), wedge-shaped,

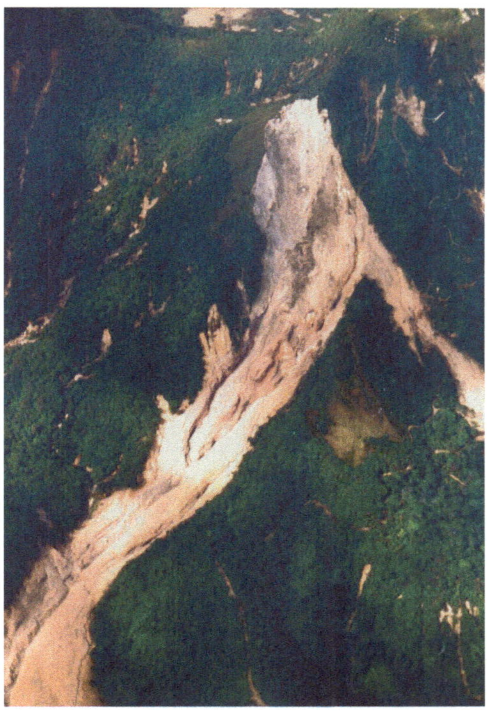

Figure 6.4 The aerial oblique shows the source area of a large debris flow on Casita Volcano, Nicaragua, as a result of heavy rainfall from hurricane Mitch on November 2, 1998. This debris flow was a catastrophic event, but note other evidence of instability in the hydrothermally altered rock of the volcano. Photograph courtesy of Molina and Hallam, 1999[4].

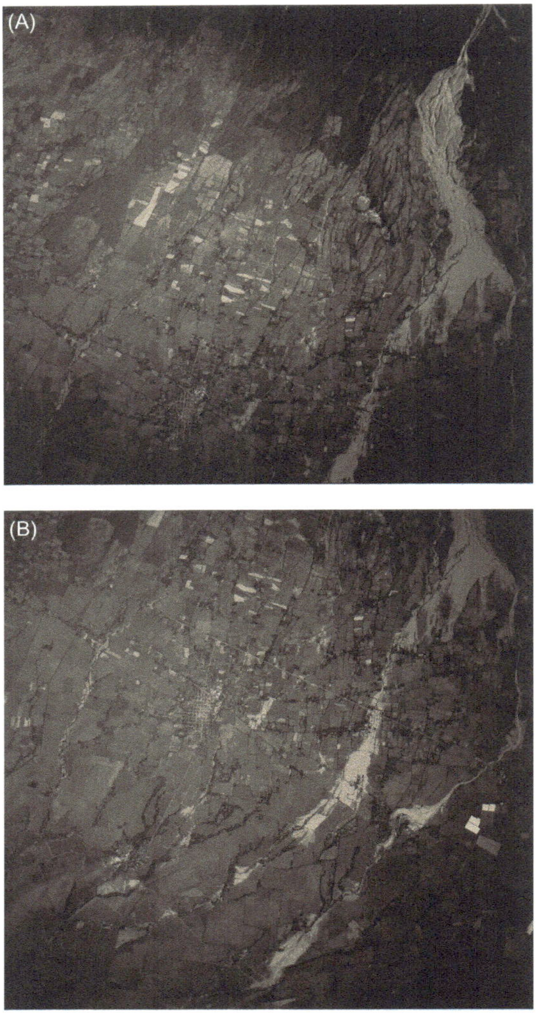

Figure 6.5 (A) The vertical aerial photograph shows the debris flow run-out of the Casita Volcano debris flow of 1998. Can you discern other evidence of previous debris flows or lahars? If so, map them. (B) This vertical aerial photograph is the stereo pair for Figure 6.5A. Photograph courtesy of Molina and Hallam, 1999[4].

where failure occurs along the intersection between two planes, or a complex combination of the two (bottom of Figure 6.7).

6.2.1 Geometry of a Rock Mass

A key to analyzing rock mass failure is achieving an understanding of the geometry of rock mass discontinuities. Aerial photography can provide help in this endeavor by delineating areas of past rock mass

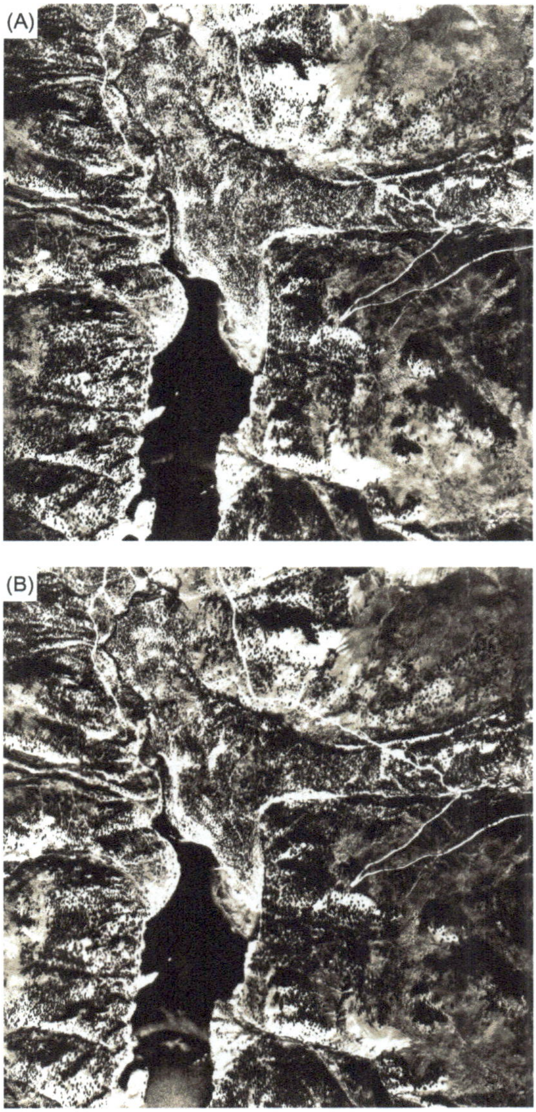

Figure 6.6 (A) The vertical aerial photograph is of a debris flow in Colorado. One of the first things to note is the presence of a lake in a steep "V"-shaped canyon. Why would one expect such an occurrence? (B) This aerial photograph is the stereo pair for Figure 6.6A.

failures as with soil slopes, but we can provide more help than that by providing quantitative, if only limited, information on the character of the blocks and discontinuities that form a rock mass.

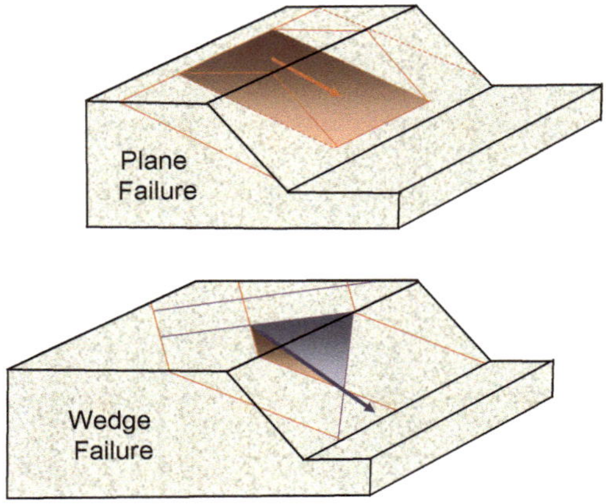

Figure 6.7 The cartoon shows plane failure geometry on top and wedge failure geometry on the bottom. Bold arrows show the direction of block movement down the failure plane or wedge intersection. Thin red and blue lines depict joints or faults. More complex combinations of joints and faults frequently occur and contribute to step-like failures.

The most common approach to characterizing the fracture pattern within a rock mass is the use of "*lower hemisphere projection plots.*" Imagine a sphere placed around the wedge failure portion of Figure 6.7. The sphere is aligned with respect to the cardinal directions. Figure 6.8 shows such an imaginary sphere. It is customary for projections drawn normal to the faces of the fractures to be projected down to the lower hemisphere and then up to the 2-D circle formed as if one looks straight down upon the lower hemisphere. In Figure 6.8, I have used the upper hemisphere for clarity to demonstrate this projection. Just remember that engineering practice usually uses a lower hemisphere projection.

Now, unlike Figure 6.8, there will be variations in strikes and dips within the rock mass fractures. So the projections of their normal lines will tend to form clusters as shown in Figure 6.9.

Depending upon photo scale, we may not be able to see all of the large fractures, or any of the small fractures, on aerial photographs. We *can*, however, use low order streams (see, for example, Figures 3.10 and 3.11) to infer fracture orientations and lengths where we can't actually see the fractures themselves. If we are lucky we may be able to obtain fracture dip from one or two fractures in each fracture set to produce an estimate for a lower hemisphere projection using aerial photographs. The odds of doing this are, unfortunately, stacked

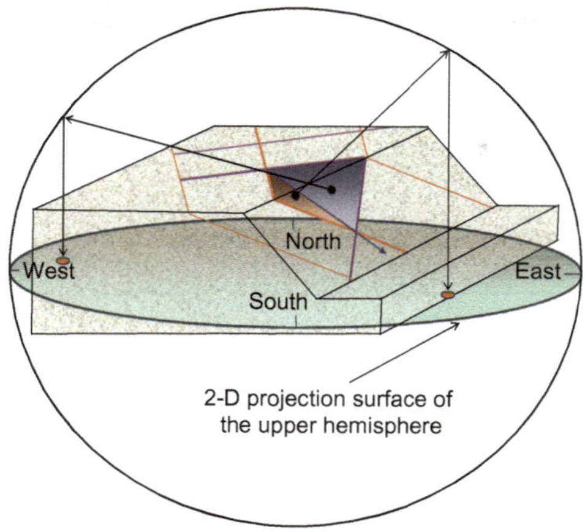

Figure 6.8 The most common approach to describing the fracture patterns within a rock mass is to project normal lines from individual fractures onto the lower hemisphere then back up to a 2-D projection of the equatorial plane. I have used an upper hemisphere projection here for clarity.

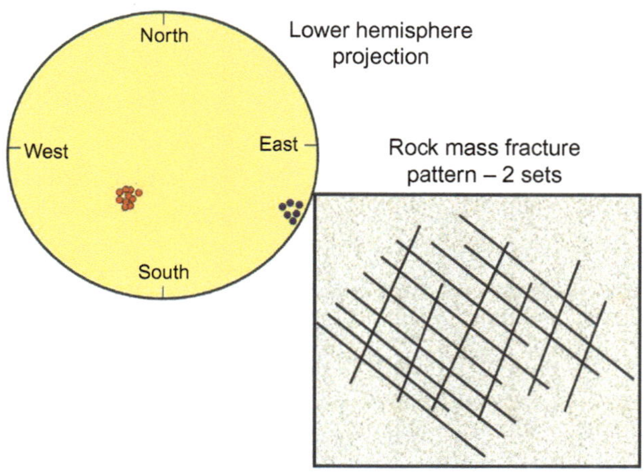

Figure 6.9 Actual rock mass fractures have variations in their individual strikes and dips. If we plot the normal lines from the fracture surfaces to a lower hemisphere projection, we will get clusters as shown here. The clusters farther toward the edge indicate steeply dipping fractures, the clusters farther toward the center indicate gently dipping fractures.

heavily against us. Try it anyway. I love it when someone gleans the maximum amount of information possible from a data set. Coarse approximations though they may be, your efforts will be appreciated.

Let us assume that our aerial photograph analysis permits us to interpret only the coarse rock mass fracture pattern shown at the lower right in Figure 6.9. Furthermore, we can find no information on fracture dips. We can still glean information that can aid in many rock mass stability studies, from rock slope stability to subsurface rock mass behavior to groundwater flow.

6.2.2 Permeability Within a Rock Masses

Now is a good time to talk a little about permeability. As a general rule, the permeability within a rock mass is controlled by flow through the fractures. There are two approaches to estimating the fluid flow through a fractured rock mass: (a) estimating the flow through discrete fractures, then combining them, and (b) estimating the flow by combining a large number of fractures together to simulate an unfractured porous media. We can employ aerial photographs for both. In the first case, aerial photographs help us determine the locations and orientations of the largest fractures. These fractures can be mapped on aerial photographs for later, more detailed, measurement and analysis. In the second case, aerial photographs can be combined with other techniques to help simulate an equivalent porous media (EPM).

To develop these two approaches look back at the rock mass fracture pattern in the lower right corner of Figure 6.9. There are two ways to look at the fracture pattern. The first is to emphasize the fractures themselves. The second is to emphasize the solid rock blocks between the fractures. Both ways will get us to pretty much the same place, but the first (concentrating on fractures) is probably more important for fluid flow, whereas the second (concentrating on rock blocks between fractures) is probably more important for rock mass slope stability. The following two sections comprise one of those mathematical diversions I mentioned previously. If you don't wish to follow them, merely skip to the final comments at the end of this chapter.

6.2.2.1 Concentrate on Fractures: The Fracture Tensor

The fracture tensor approach was first formulated by Oda[2]. Assuming there are m fracture sets, the fracture tensor in 3-D is defined as

$$F_{ij}^{(m)} = 2\pi\rho \int_0^{\infty} \iint_{\frac{\Omega}{2}} r^3 n_i n_j f(n,r) \mathrm{d}\Omega \, \mathrm{d}r$$

In the above equation, ρ is the fracture density (number per unit volume), r is the radius of a circular fracture, n_i and n_j are the outward normal components in the i and j directions, $f(n,r)$ is the joint probability density function for n and r, and $\Omega/2$ is the solid angle corresponding to the surface of a unit hemisphere (radius = 1). If the fracture radius is not correlated with the fracture orientation, the joint probability density function, $f(n,r)$ can be uncoupled so that $f(n,r) = f(n) f(r)$. In this case and in 2-D, the fracture tensor becomes

$$F_{ij}^{(m)} = \rho \int_0^{r_{\max}} r^2 f(r) \mathrm{d}r \int_0^{\pi} n_i n_j f(\theta) \mathrm{d}\theta$$

The first integral is simply the histogram moment of order 2 of the histogram of fracture set radii. The second integral sums over fracture orientations, and θ is the angle between the fracture direction and the x-axis. The histogram moment of order 2 about the origin is given by

$$M_2(r) = \int_0^{\infty} r^2 f(r) \mathrm{d}r = \frac{1}{N} \sum_{r=0}^{L-1} r^2 N(r)$$

If we wish to know the fracture tensor in some specific direction, p, the directional fracture tensor component for fracture set m is

$$F_p^{(m)} = \rho^{(m)} \left[\frac{1}{N} \sum_{r=0}^{r_{\max}} r^2 N(r) \right] \int_0^{\pi} n_p^2 f(\theta) \mathrm{d}\theta$$

In the above equation, the integral is over fracture orientations, and n_p is the projection of a fracture at angle θ_i onto vector p. This projection is given by $n_p^2 = \{||1|| \, ||p|| \cos \alpha\}^2$, where α is the angle between θ and p, and $||p||$ is the length of p. Note that the integral is the second histogram moment about the origin of angles, so it becomes

$$\frac{1}{N} \sum_{\theta=0}^{\pi} (||1||||p|| \cos \alpha)^2 N(\theta)$$

and the final equation for the fracture tensor component in the p direction for fracture set m is

$$F_p^{(m)} = \frac{\rho}{N} \left\{ \sum_{r=0}^{r_{\max}} r^2 N(r) \sum_{\theta=0}^{\pi} (||1||||p|| \cos \alpha)^2 N(\theta) \right\}$$

We want the fracture tensor component for all sets of fractures, this will be

$$F_p = \sum_{m=1}^{N} F_p^{(m)}.$$

We will choose the direction p with some feature of interest in mind, such as excavation support, permeability, and free rock face. So we see that the fracture tensor is nothing more than a statistical quantity computed using histograms. This can be visualized by looking at Figure 6.10.

I have put you through all of this derivation so that, if you care, you can program the above equations and solve for the fracture tensor yourself. I chose Mathcad for programming rather than some other software like Matlab, because I find it more visually instructive. You can easily program it in C, C++, Fortran, Matlab, or some other software program. I have run Mathcad for the fracture pattern at the lower right of Figure 6.9 for several p directions and graphed it in a polar plot to make it easier to visualize the fracture tensor (left polar plot of Figure 6.11). The fracture tensor is dominated by the longer fractures in the northwest–southeast direction, as you would expect. Many authors have demonstrated a correlation between the fracture tensor and the permeability tensor in 2-D, so a plot such as this can also provide insight into the permeability tensor (the direction that subsurface water will move). I have cleverly restricted consideration to

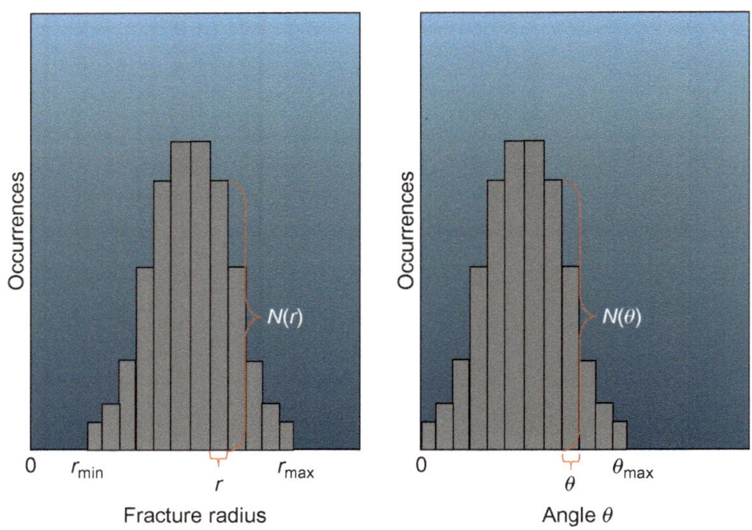

Figure 6.10 The histograms demonstrate the statistical nature of the fracture tensor component. I have constructed the histograms in the shape of a Gaussian distribution, but since we directly use the entire histogram without assuming a functional distribution, the computation of the fracture tensor is nonparametric, thus it does not depend on the shapes of the histograms.

2-D. This favors an analysis using aerial photographs and eliminates a dependence on fracture dip needed in 3-D.

6.2.2.2 Concentrate on Rock Blocks: The Fry Diagram

The Fry diagram[3] is a center-to-center spatial analysis designed by Fry to represent nonisotropy in mineral crystals as an indicator of strain. Fry's contention was that crystals growing within a melt should be randomly isotropic in shape. Stress applied later to the rock would tend to elongate the crystals in a preferred strain direction. The contention demands a correlation between strain and crystal elongation. When we deal with fractured rock masses, we need no such relationship as long as we do not attempt to infer a stress or strain direction directly from the anisotropy of rock blocks. All we wish to do is provide some indication of the directions at which rock blocks are larger or smaller. The Fry approach is best applied using 300 or more rock blocks. The fracture pattern of Figure 6.9 has only 25 rock blocks.

Both the Fry analysis and the fracture tensor analysis are statistical descriptions of a rock mass. The fry analysis displays two-thirds of the information in the fracture tensor, spacing (block size), and orientation. The missing component is the fracture radius (fracture length).

6.3 SOME FINAL COMMENTS ON LANDSLIDES

First, remember that it doesn't take much movement along a slide plane to force groundwater to the surface. Landslides do not always

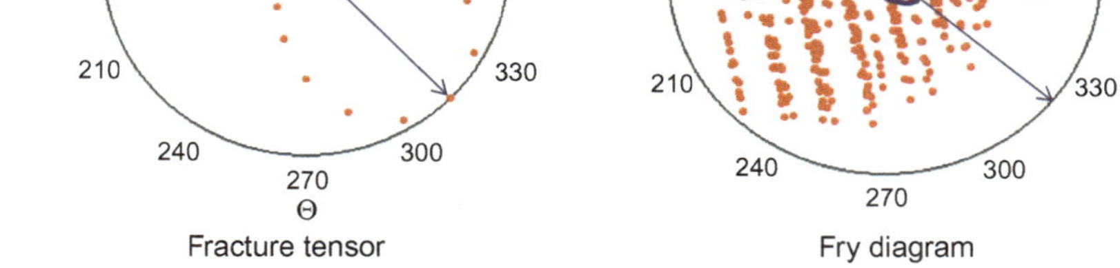

Figure 6.11 These polar plots display the results of a fracture tensor analysis on the left, and a Fry analysis on the right. Both analyses were done using the fracture network shown in the lower right of Figure 6.9.

fail catastrophically as soon as resisting forces are exceeded. Some landslides creep slowly before they fail. Insipient landslides may sometimes be detected by seeps at the sloping rock or soil face. Always pay attention to springs or seeps on hillsides as these may forecast landslide dangers to come.

Second, as I mentioned earlier, if you build on a landslide that you should have seen on an aerial photograph, expect to end up in court with someone like me sitting across from you with an enlarged aerial photograph showing *your* landslide. That's right, *your* landslide. You may as well give it your name, because you are about to buy it. I know of engineering firms that have created real estate/development companies to buy their landslides. They would later stabilize and resell them.

Third, and finally, I have had developers tell me, "We know of the landslide, but it is an old landslide, one that occurred during more humid conditions, and is now stable under present drier conditions." This all may be true, but the developer is about to significantly alter the present "drier" conditions. He is going to add building loads, swimming pools (which always leak), irrigation systems to make drier conditions more aesthetic (reestablishing the more humid conditions), and road networks to funnel sparse rainfall quickly into his landslide. It has now become his landslide, because I guarantee he will soon be buying it if he doesn't already own it.

REFERENCES

1. Glass CE, Klimmek R. Routing debris flows. Environ Eng Geosci 2001;VII(2):177–91.
2. Oda M. Fabric tensor for discontinuous geological materials. Soils Found 1982;22:96–108.
3. Fry N. Random point distributions and strain measurement in rocks. Tectonophysics 1979;60:89–105.
4. Molina, B.F. and C.A. Hallam. Open Skies aerial photography of selected areas in Central America affected by Hurricane Mitch 1999, USGS Circular 1181.

CHAPTER 7

Dangers from Floods

Floods in Pakistan during the 2010 annual monsoon season underscore the importance of flood management as a worldwide imperative. Floods there killed more than 1500 people and stranded hundreds of thousands. The United Nations estimated that as many as 21 million people were injured or displaced[1], and at one time as much as 20 per cent of the total land area of Pakistan was under water[2].

In the United States, historically flood damage has steadily increased despite local and federal mitigation efforts[3]. Although there are no uniform guidelines for estimating flood losses, and no central clearing house exists for data on flood damage, estimates are that floods caused as much as $2.4 billion in damage in the United States in 2003, and more than $16 billion in 1993, the worst year on record. Increases in damage could be the result of a number of things, including: (a) population increases in flood-prone areas, (b) climate change, and (c) development patterns and local flood policies. One thing is certain; flood damage hazard is directly proportional to proximity to streams, rivers, deltas, and coastlines. If we supplement available stream flow data with a careful study of the geomorphology of streams, rivers, deltas, and coastlines within an area, we can better estimate flood hazard.

Deltas, for example, may be prograding or retrograding. High sediment discharge rates in prograding deltas favor levee longevity as levees become positioned farther inland, and hence are stabilized naturally or artificially by engineers. Unless artificially supported, levees in retrograde deltas tend to be transient and less stable.

An additional factor to consider is coastline morphology relative to ocean currents, tides, and weather patterns (typhoons and hurricanes, for example).

7.1 STREAM FLOW DATA AND THE MYTH OF THE 100 YEAR FLOOD

Even though stream flow data are not readily available from studying aerial photographs, they are important to the overall understanding of the risk due to floods in any particular area. The USGS provides stream gauge data for most major streams in the United States. The data are usually presented in the form of stream flood hydrographs as shown in Figure 7.1.

Hydrologists have observed that all hydrographs have the same basic shape; they possess a rising limb, a crest, and a receding limb. The area under the hydrograph is the total discharge from the storm (remember to account for proper units). The similarity of hydrographs has led to the concept of a *unit hydrograph.* A unit hydrograph is a discharge hydrograph resulting from one unit of net precipitation distributed uniformly over a watershed. Hydrographs are relatively shape invariant. We can scale the unit hydrograph to account for different precipitation intensities and durations, providing insight into flood behavior for individual watersheds. Care must be exercised in using such simple models for prediction purposes. It is assumed, for example, that rainfall is uniform over the watershed. This assumption breaks down for convective storms that may exhibit intense rainfall over small portions of a watershed, as frequently occurs in the southwestern United States. Nevertheless, I have always had a great appreciation for simple models. They should always be used with care, keeping in mind the simple underlying assumptions, but they

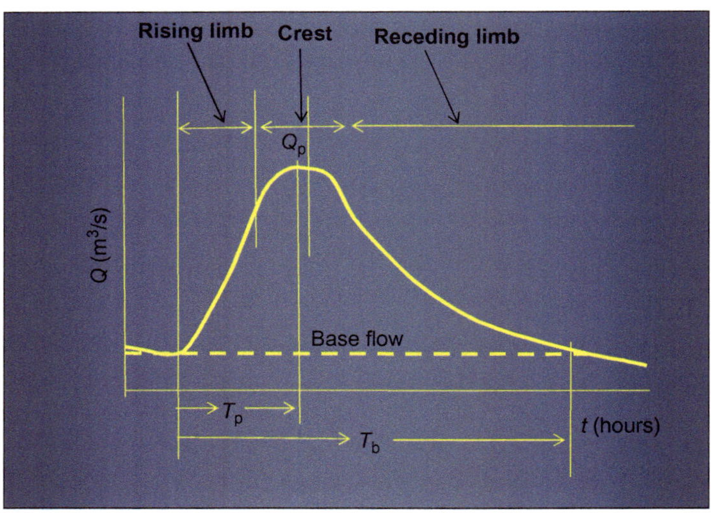

Figure 7.1 The graph shows a typical flood hydrograph. The vertical axis is the stream discharge in cubic meters per second, and the horizontal axis is time in hours or days. Q_p *is the peak flood discharge,* T_p *is the time to peak discharge, and* T_b *is the time from flood onset to a base flow condition.*

usually provide important, if not necessarily precise, insight into the process in which we are interested.

More complex hydrological models attempt to route floods downstream to assess time-varying peak flows. To do this accurately, we need detailed information on stream and watershed geometries, infiltration models, and rainfall models.

In many cases, the only gauging information available to us is an annual maximum value for Q_p, the peak flood discharge for a given flood event. Typically, one will rank the peak annual discharges from maximum to minimum assigning a rank to each value, from 1 (maximum annual discharge) to min (minimum annual discharge). The average return period can then be computed for each stream gauge data point as

$$T = \frac{n+1}{m}$$

In the above equation, T is the return period in years, n is the number of years of stream gauge data, and m is the rank of each annual maximum flood. An example is provided in Table 7.1.

If we now plot the stream gauge maximum annual discharge from Table 3.1 against the return period, T, we get Figure 7.2. Obviously, since we have only 25 years of stream gauge data we must extrapolate (guess) to estimate a 100 year return period. Why do we care about the 100 year return period? The 100 year flood is defined as that flood having an average return period of 100 years. Why, then, is the 100 year flood important?

In 1973, the National Flood Insurance Program undertook a project to set standards for mapping flood-prone areas for issuing regulations, development standards, and provide flood insurance guidelines. The National Flood Insurance Program adopted the 100 year flood as its mapping standard for designating floodplains. Banks and other lenders quickly adopted the 100 year floodplain maps to guide their requirements for the purchase of flood insurance. Since then, the term "*100 year flood*" is almost universally used to describe any significant flood event regardless of recurrence probability. So what is the nature or the myth?

In Figure 7.2, we can see that there are an infinite number of paths (I have illustrated only two—dashed lines) from our 25 year data set out to 100 years. So which of these paths provide us with the true 100 year

Table 7.1 Example of Stream Gauge Data for a Period of 25 Years

Rank	Discharge (m^3/s)	Return Period (Years)
1	380	26
2	375	13
3	370	8.666667
4	365	6.5
5	351	5.2
6	339	4.333333
7	323	3.714286
8	308	3.25
9	300	2.888889
10	287	2.6
11	277	2.363636
12	263	2.166667
13	255	2
14	233	1.857143
15	231	1.733333
16	222	1.625
17	220	1.529412
18	210	1.444444
19	199	1.368421
20	196	1.3
21	183	1.238095
22	172	1.181818
23	162	1.130435
24	153	1.083333
25	145	1.04

flood? Is the 100 year flood 580 m^3/s or nearer to 800 m^3/s? As the stream gauge data set includes more years of information, the definition of the 100 year flood changes. If, for example, the 26th year produces a 580 m^3/s flood, a new graph projecting from the new datum will predict a much more severe 100 year flood. So, do we keep the old value and figure we have just experienced the 100 year flood, or do we use a new one? Herein lies the myth of the 100 year flood.

There is another problem with relying on this concept of a 100 year flood. As it is a purely statistical quantity, it bears scant relationship to

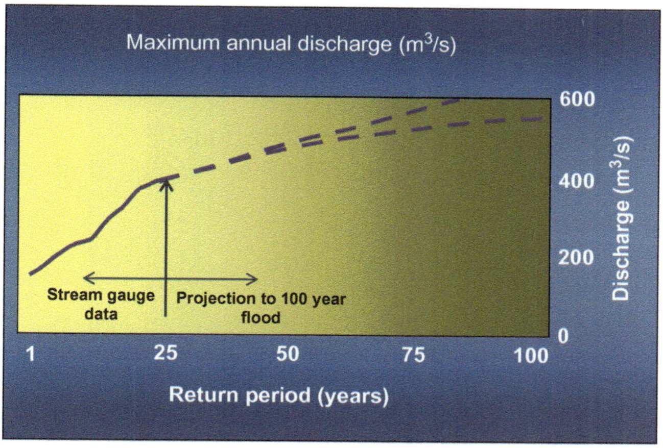

Figure 7.2 The plot of the data in Table 3.1 extrapolated to a 100 year return period.

the watershed in question—can the watershed even support that level of flood? Perhaps the trajectory in Figure 7.2 that flattens with time is more reasonable. Perhaps it is not. *If we wish to better understand low probability events, we should go to the source of the low probability events, not merely extrapolate from high probability events.* Here is where aerial photographs supported by ground studies can help us.

Let us look at two different flooding scenarios. First, consider large, perennial rivers in relatively humid environments. Geomorphologies of large perennial river systems develop over thousands of years of periodic flooding and inter-flood processes. Characteristic landforms to look for on aerial photographs include:

- Undercut banks.
- Point bar deposits.
- Natural levees.
- Meander scars.
- Cutoff meanders.
- Extensive floodplains.
- Crevasse splay deposits—subsidiary lobate deposits of coarser-grained sediment that accumulate along the landward sides of channel levees in area where levee breaches occur.
- Slack-water deposits—deposits of fine-grained sediment that accumulate near the channel in areas of abrupt decrease in flow velocity during flood events. Areas of ineffective flow during extreme floods

are the most conducive settings for accumulation of slack-water deposits. Slack-water deposits are found in the following areas:

- Back-flooded tributary mouths.
- Areas downstream of channel obstructions.
- Eddy areas associated with abrupt channel contractions and expansions.
- Broad floodplain areas.

- Erosion marks.
- Tree scars.
- Debris lines.

Factors tending to increase flood magnitude include:

- Large drainage areas.
- Steep basins.
- Lack of vegetation cover.
- High altitude basins, which can release stored water rapidly and unexpectedly.
- Watersheds with round shape.
- Shallow depth to relatively impermeable strata (e.g., igneous rocks and shale).

As a general rule of thumb, wide, shallow riverbeds indicate instability in flow through channels. Friction of the channel to water flow can be assessed by viewing the meander length. A low-friction channel, for example, produces long wavelength meander patterns and is indicative of high flood flows, a potential danger to nearby engineering projects. Low-friction channels move water through faster, contracting the flood hydrograph. Some observations of meander wavelength as a function of stream discharge are presented in Table 7.2.

Channels with more cohesive materials (shale, volcanic ash, loess, and bedrock) tend to be relatively narrow, deep, sinuous, and have smaller wavelengths. Larger wavelengths are associated with higher proportions of bed-load transport.

Meander migration rate has an important affect on structures built downstream of the major swing of a meander. Migration rate is a

Table 7.2 Discharge as Related to Meander Characteristics

Source	Relationship	Comments
Dury[4]	$\lambda = 54Q^{0.5}$	Q = mean annual flood
Carlston[5]	$\lambda = 166Q^{0.46}$	Q = mean annual discharge
	$\lambda = 126Q^{0.46}$	Q = mean monthly maximum discharge
Ackers and Charlton[6]	$\lambda = 62Q^{0.47}$	Q = constant bank-full discharge in lab streams
Dury[7]	$\lambda = 33Q^{0.55}$	Q = most probable annual flood

maximum when the radius of curvature (r_c)/stream width (w) corresponds roughly to the following equation

$$2 \leq \frac{r_c}{w} \leq 3$$

Second, consider intermittent rivers in relatively arid environments. Flood indicators in these areas may include:

- Scour lines
- Silt lines
- Flood debris
- Flood-damaged vegetation
- Terraces.

The response of intermittent streams to excessive water volume is usually to increase the channel by scouring (to deepen the channel) and eroding or undercutting the lower parts of exposed banks (to widen the channel). This is a common occurrence in arid to semiarid regions having weak riverbank material, and unless bank protection is undertaken, over-bank flow and floodplain deposits are rare. Any structures built on the riverbank, or anchored there (such as bridges), are susceptible to damage due to this type of channel enlarging. Local channel protection strategies commonly employed to protect structures on, or anchored to, riverbanks accomplish two things. First, they protect the channel banks from erosion, thus guaranteeing over-bank flow. Second, they decrease the friction of the channel, speeding the flood downstream and altering the flood hydrograph. Anything you do to a riverbank has an effect, usually a detrimental one, on downstream populations. Landforms common to intermittent streams include:

- Multiple terraces
- Cutoff meanders
- Wide, shallow beds.

7.2 ALWAYS GIVE SOME THOUGHT TO THE POSSIBILITY OF CATASTROPHIC FLOOD EVENTS

In the early 1800s, the study of geology was burdened by conflicts between two camps. Geologic scientists believed that the same natural processes that operate and affect the earth today have always operated in the past and affected the earth in the same way. Biblical scholars believed that the distant past "*consisted of epochs of paroxysmal and catastrophic action interposed between periods of comparative tranquility*"[8] (Noah's flood is a good example).

Names for the scientific approach and the biblical approach were coined by William J. Whewell as *uniformitarianism* and *catastrophism*, respectively. Through the intervening years, geologists became averse to admitting that catastrophic events had a role to play in geologic processes for fear of receding into biblically influenced ideology.

Then in the twentieth century a debate was centered on the Channeled Scablands of the Columbia Plateau in eastern Washington (Figure 7.3).The debate was between geologists who insisted that the Scablands were the result of millions of years of erosion conforming to uniformitarianism, whereas other geologists insisted that they were the result of catastrophic floods during the Pleistocene generated by periodic rupture of glacial moraines impounding glacial Lake Missoula in Montana[9].

Evidence finally became overwhelming that periodic rupture of glacial moraines sent huge floods from Lake Missoula over the Columbia Plateau scouring deep gorges (Coulees) into the basalt. Geologists still believe in uniformitarianism, but nevertheless understand that periodic catastrophes are also an important part of the process. So when dealing with published equations and models for flood hazard estimation, stream meander wavelength versus discharge volumes, unit hydrographs, etc., keep in mind that an event that is yet unknown historically

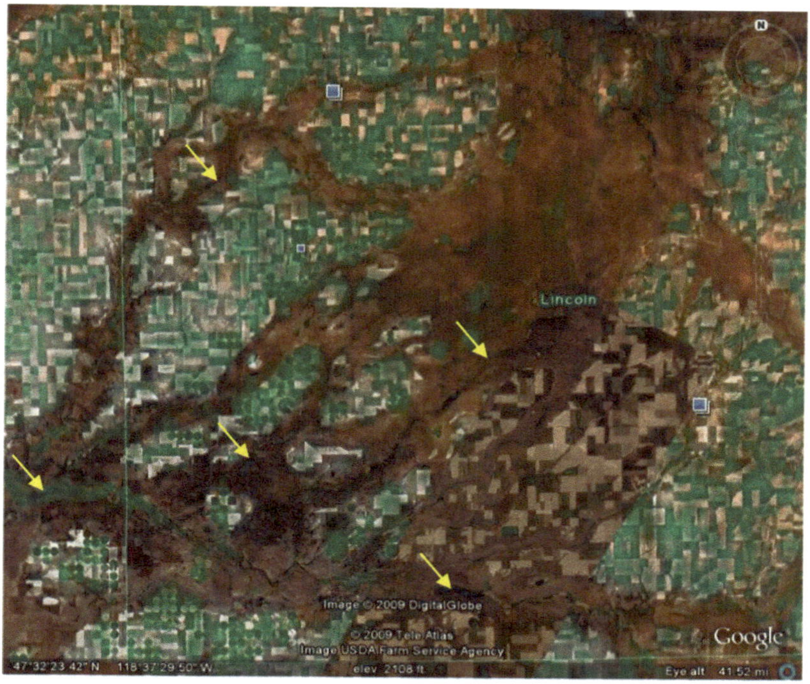

Figure 7.3 The vertical view is a Google Earth image composite of the Channeled Scablands of eastern Washington. The deep gorges (Coulees) were formed by periodic catastrophic floods from Lake Missoula.

may be just around the corner to humble the most confident geologist, engineer, and planner. One such event occurred in southeastern Arizona on October 10, 1983.

On October 10, 1983 Tropical Storm Octave progressed north along the Sea of Cortez between Sonora and Baja, Mexico and stalled over the Santa Cruz River watershed in northern Sonora, Mexico and southeastern Arizona. The resulting 23 cm (9 in.) of widespread rainfall added to 18 cm (7 in.) of antecedent rainfall in August and September 1983 was responsible for an epic flood that killed four people and caused more than $100M damage in 1984 dollars. The flood is still discussed in reverent tones in Arizona. Figure 7.4 shows a vertical aerial photograph of a portion of the Santa Cruz River south of Tucson near Tubac, Arizona after the 1983 flood waters had returned to near the base flow, well out on the receding limb of the flood hydrograph.

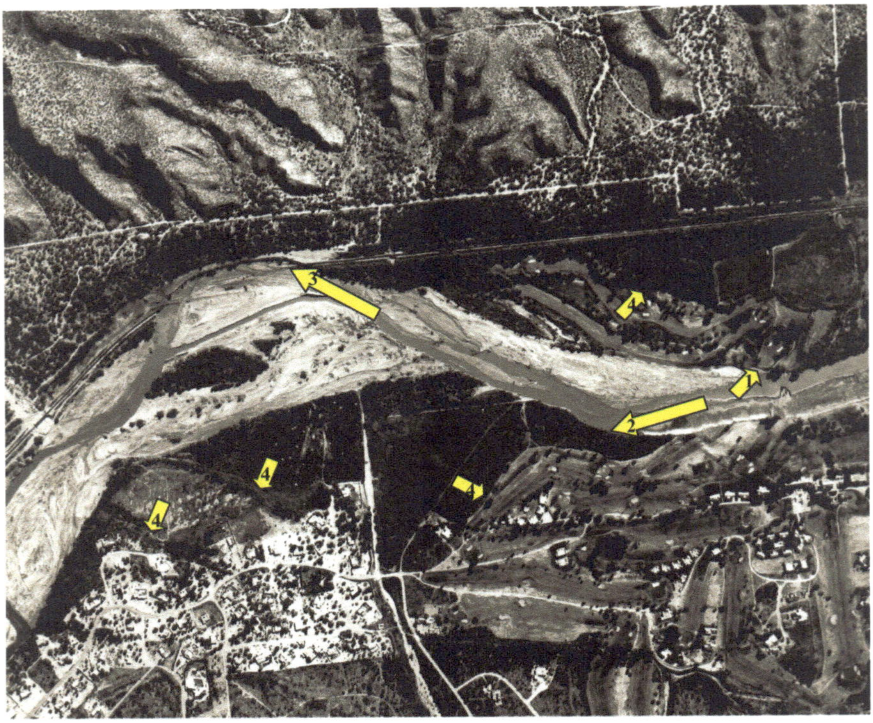

Figure 7.4 The vertical aerial photograph shows the Santa Cruz River near Tubac, Arizona soon after the epic 1983 flood. North is to the left. The Santa Cruz River is one of the few rivers in the United States that flows northward.

Flood features of note in Figure 7.4 are delineated by numbered yellow arrows. Arrow number 1 points to bank erosion on the east edge of the Santa Cruz River. Similar erosion can be seen on the west bank. Bed and bank erosion has cut a golf course in half. Note the relatively straight course of the river to the south (right). This straight segment propelled the flood waters straight at a meander (arrow 2). Flood water eroded the soft bank for a distance approximately equal to the length of arrow 2, until it was forced to the east by the increased resistance due to the land. The east bank was then subjected to strong erosional forces and yielded, again approximately the distance of arrow 3, by widening the river channel. The Southern Pacific Railroad bed was taken downstream to the north. Note the shorter wavelength meander favored by the Santa Cruz River under lower, base flow conditions. The narrow straight segment of flow near arrow 3 was excavated to divert flow away from repair operations to the railroad bed. There was no over-bank flow during this event, as the river widened to accommodate the increased flow. Probable previous erosion scars can be seen at yellow arrows 4.

The volume of flow during this extreme event was historically unprecedented. Could such an event have been foreseen? Possibly; I just don't know. Careful mapping of past erosion scars (arrows 4 on Figure 7.4, for example) utilizing aerial photographs may indeed provide insight into prehistoric events of comparable magnitude. Dating of geomorphic surfaces associated with prehistoric events may also provide an idea of the return period for large events. Erosion and avulsion of river systems frequently occur asymmetrically, however, removing terraces from one bank relative to the other and making dating and matching of terraces difficult. Studies should integrate information along enough of a river's reach to provide confidence in the information.

My point here is that we, as scientists, engineers, resource and public managers, and planners should not become complacent merely because available statistical and physical models appear to be good estimators under current conditions. *The present may not always be a key to the past,* ***or to the future***. Look carefully at the entire watershed system using all of the aerial and ground-based tools at your disposal. Don't become narrowly focused on your particular site, either. Upstream conditions, natural or manmade, may have a profound impact on your project.

For example, dams are engineering structures especially susceptible to unexpected large floods. It is common during large floods for dams on the main water course as well as on subsidiary water courses to fail. Watch out for upstream dams of all kinds, from flood control dams to smaller retention dams. Their failure can profoundly and disastrously alter the flood hydrograph at your site. Other structures that intersect or lie adjacent to a river course (highway embankments, train or road overpasses, mine tailings, etc. are examples) can be identified on air photos, as well as occasionally on topographic and urban maps. Failures of downstream features or dams also may significantly alter flood hydrographs. Pay attention to all of these.

It is often useful to review the status and quality (type, density, etc.) of vegetation that occurs upstream of potentially human impacted areas (e.g., an upland woodlands) and to access the potential buffering effect of these on flood hydrographs. Remember, though, that such features are notoriously ephemeral. They may change with population stress, climate, and biological factors.

REFERENCES

1. South Asia BBC News (8/14/2010), Floods affect 20 million people—Pakistan PM Gilani, BBC.
2. Goodwin L. One fifth of Pakistan underwater as flooding disaster continues, 2010. News. Yahoo.com.
3. Pielke Jr. RA, Downton MW, Barnard Miller JZ. Flood damage in the United States 1926–2000: a reanalysis of National Weather Service estimates. Boulder, CO: UCAR; 2002.
4. Dury GH. Principles of underfit streams, USGS Prof. Paper 452-A. 1964.
5. Carlston CW. The relation of free meander geometry to stream discharge and its geomorphic implications. Am J Sci 1965;263:864–85.
6. Ackers P, Charlton FG. The slope and resistance of small meandering channels, Proceeding of the Inst. of Civil Engrs., Paper 7362S, Supplement XV, 1970. p. 349–70.
7. Dury GH. The face of the earth. Pelican Books Penguin, Penguin Books LTD, UK; 1977.
8. Whewell WJ. Principles of geology, Charles Leyell, Vol. II, London, Quarterly Review, 1832;47:103–23.
9. Bretz JH. The Channeled Scabland of the Columbia Plateau. J Geol 1923;31:617–49.

CHAPTER 8

Dangers from Ground Subsidence

There are a number of artificial and natural causes for ground subsidence. In the southeastern United States, for example, ground subsidence occurs as artificial and naturally induced fluid migrates in relation to the dissolution and removal of evaporite deposits. Farther west, natural plastic deformation and migration of salt within formations (the Paradox Basin is an example) can cause ground subsidence. Also in Utah, NM and many regions in the Mid-west and Appalachian area, underground coal seams and mines that are relatively rich in methane naturally ignite as oxygen is introduced through erosion or mining. Short- and long-term conversion of underground burning coals to clinker results in a significant volume loss that can lead to subsidence. In other mining areas, trimming of mine pillars can cause their failure, inducing subsidence of overlying geologic materials. Withdrawals of petroleum in many sedimentary basin units (deeper consolidated units or shallower and/or young, overpressured, water-rich, and unconsolidated units) results in surface subsidence when proper reservoir pressure management (e.g., secondary water-flood or tertiary fluid pressure support) is not applied.

In the southwestern United States, groundwater withdrawal and collapsing soils generate ground subsidence and fissures. In the Gulf Coast, the prograding Mississippi Delta causes "growth fissures" as landward sediment deposition causes instability at the toe of the delta.

Ground subsidence per se is not a serious hazard unless subsurface infrastructure (pipelines, tunnels, etc.) is adversely impacted. Tensile stresses induced at the margins of the subsidence area, however, eventually cause ground failure to develop in the form of fissures at the ground surface. I will discuss the latter three causes of surface fissures.

8.1 GROUND SUBSIDENCE DUE TO GROUNDWATER WITHDRAWAL

Ground subsidence hazard due to groundwater withdrawal is somewhat unique to the Southwest. It occurs when groundwater is withdrawn from a basin aquifer or subsurface reservoir faster than natural or artificial recharge can replace it. In the case of shallow groundwater aquifers in the Southwest, rapid withdrawal of groundwater causes regional groundwater levels in a basin to decline. Accompanying groundwater level decline is a decrease in pore pressure. A decrease in pore pressure causes an increase in effective stress, which causes the sediment to compress. Sediment compression eventually results in surface subsidence and finally subsidence fissures. Ground subsidence in some sedimentary basins in Arizona, for example, exceeds 5 m.

Ground subsidence is not usually discernable on aerial photographs, because it often occurs over broad areas. Subsidence fissures, however, occur along the margins of basins where the tensional effects of ground subsidence are severe. These fissures are easily seen on aerial photographs. Subsidence fissures may alter natural drainage, damage roads and other lifelines such as fuel and power conduits, provide access to groundwater resources for pollution, and damage commercial or residential structures. Recognizing them is important.

Subsidence fissures begin at the margins of subsiding basins, probably as a small crack at the ground surface. Unfortunately, to my knowledge anyway, nobody has ever witnessed this formative event. That's probably because water finds the incipient cracks first and quickly begins to erode them, making them wider and deeper. Although subsidence fissures begin at basin margins, they do not necessarily form concentric contours around the basin as one might expect. Local soil type and structure, surface and subsurface geometry, and subsidence rate probably contribute significantly to eventual fissure shape, length, and distribution throughout the basin.

Young fissures are tensile in nature as seen on Figure 8.1, hence they usually display little or no vertical or horizontal offset across the fissure. Young fissures look more like a simple crack than a fault along which shearing has removed most irregularities. Subsidence fissures tend to be highly irregular, both in plan view and in cross section. Because streams quickly occupy them, many unaware state or local transportation personnel, land use managers, and real-estate developers confuse them with dry washes, subsequently placing infrastructure entities over them with conduits to permit water flow beneath. They don't realize that water seldom flows along the fissure for long as most fissures form nearly parallel to basin contours, or trend uphill. Water quickly drains vertically into the

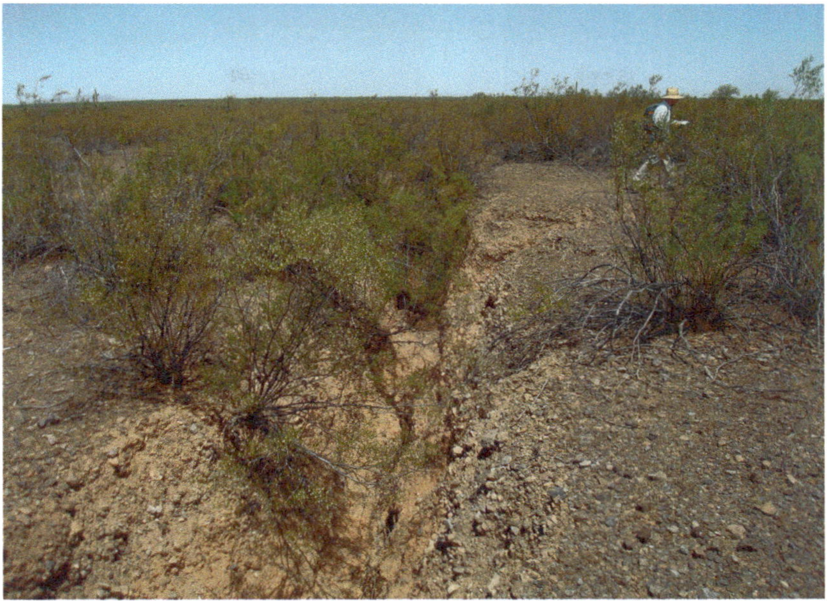

Figure 8.1 A view west at a subsidence fissure north of Picacho Peak State Park, Arizona. Photo courtesy of Dr. Robert R. Casavant.

fissure forming a rapid shortcut into the groundwater table for surface water and pollutants.

As the fissures or fissure swarms grow in extent and depth they intercept and capture small streams. Where these streams enter the fissure, they erode small "*sink holes*" as water is funneled into the fissure. The irregular crack and periodic "sink holes" form a pattern somewhat similar to a string of prayer beads. In Figure 8.1, you can see where water has formed one of these "*sink holes*" immediately to the left of the individual.

Fissures are lengthened and widened by erosion, dissolution, and removal of supporting sediments, together with weak inter-grain cementation along the fissure. This often occurs with significant apparent deformation or erosion of overlying near-surface or surface sedimentary layers. This "piping" process occurs from the bottom upward and is often masked by a "bridge" of intact surface layers that either filled a shallow and incipient fissure or have yet to be deformed. Piping is being evaluated as a favorable model for fissure formation. This hidden process makes incipient fissure formation difficult to interpret on aerial photos as well as during field investigations[1].

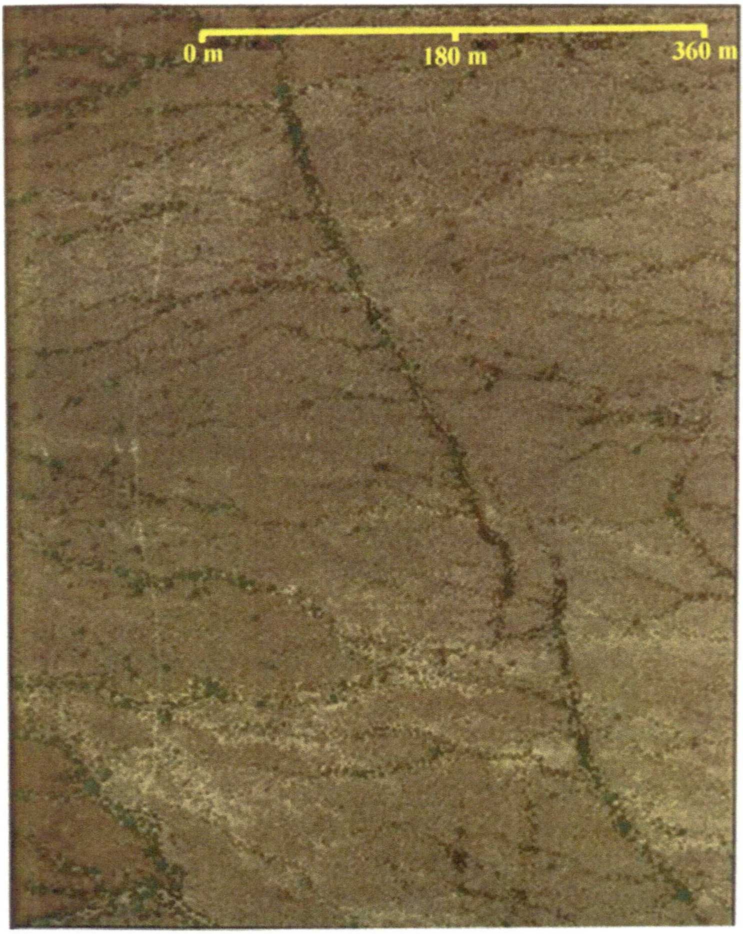

Figure 8.2 The oblique aerial photograph shows several subsidence fissures north of Picacho Peak State Park, Arizona. Note the irregular pattern and preferential vegetal growth. Note also that these fissures trend roughly perpendicular to regional stream flow directions and have captured several streams. Photo courtesy of Dr. Robert R. Casavant.

As subsidence fissures mature, vegetation (creosote bush and mesquite are common in Arizona) grow preferentially along the fissure, taking due advantage of water channeled there. As the fissure widens and lengthens, it captures more streams, thus promoting even further growth. Mature fissures may also display differential offset across the fissure, abrupt terminations, and offset fissure segments. Figure 8.2 shows an aerial view of subsidence fissures north of Picacho Peak State Park, southeast Arizona.

As subsidence fissures continue to enlarge, they collect more than water and sediment, frequently becoming depositories for household and industrial garbage, hazardous materials, unsuspecting animals, and perhaps even children. Although the large fissure shown in Figure 8.3 has no such detritus, other than the two geologists standing there, you can see how accidents can happen. In a large subsidence fissure near Chandler, Arizona my students and I found a small puppy that had fallen in. It took the compassion of my students together with expertise and equipment of the local Rural Metro Fire Department to free the dog.

8.2 GROUND SUBSIDENCE DUE TO COLLAPSING SOILS

Collapsing soils are another phenomenon common to the Southwest. Collapsing soils occur when flood deposits form young terraces adjacent to desert washes. Silt and clay are deposited rapidly forming a *flocculated* soil structure as shown in Figure 8.4.

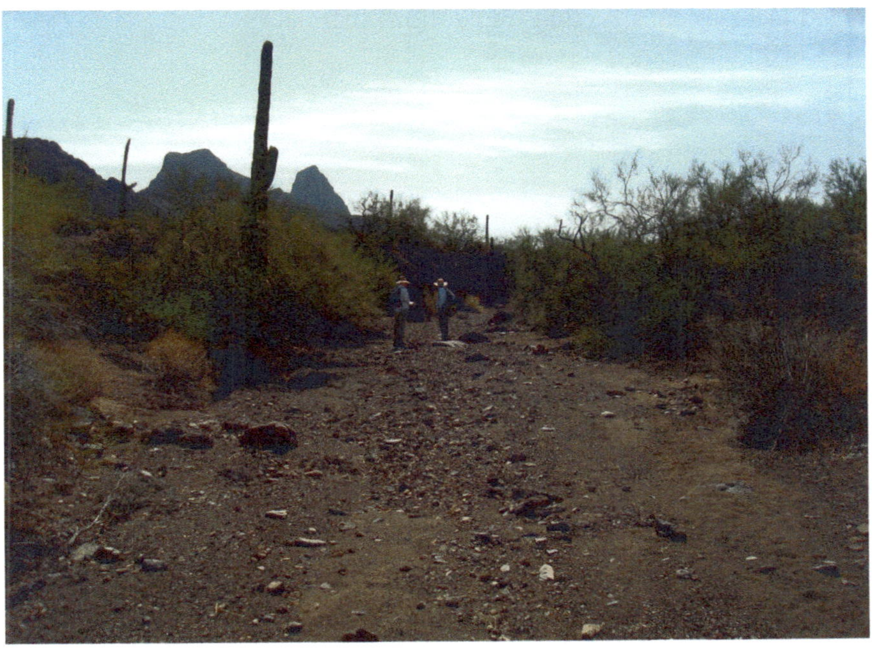

Figure 8.3 The photograph facing west shows Picacho Peak in the distance, and a large subsidence fissure in the foreground. The fissure walls are easily 3 m high here. You can see how state transportation personnel could mistake the fissure for a stream channel.

The flocculated structure is held together by salt, calcium carbonate, or silt bonds between the clay platelets. These youthful deposits are highly under-consolidated, so when subsequent construction occurs followed perhaps by irrigation, the bonds holding the flocculated structure in place are dissolved, causing the soil to consolidate into a more natural dispersed structure (right in Figure 8.4). This consolidation can cause differential settlement of up to a meter. Structures built atop such sediment can experience serious damage. Although the soil structures that cause a soil to collapse are not visible on aerial photographs, the geomorphology related to their occurrence is (Figure 8.5).

Figure 8.5, the same aerial photograph we used to look at flood geomorphology displays several geomorphic features common to areas conducive to collapsing soils. Areas having potentially collapsing soil hazard can be delineated by marking the youngest terraces (red stars). These young terraces warrant soil testing to assess collapse potential of soils. Both the railroad and the apparent subdivision (westernmost red star) have relatively high collapse potential. Older terraces (green stars) probably have normally consolidated soils, but tests should be conducted prior to constructing foundations there also.

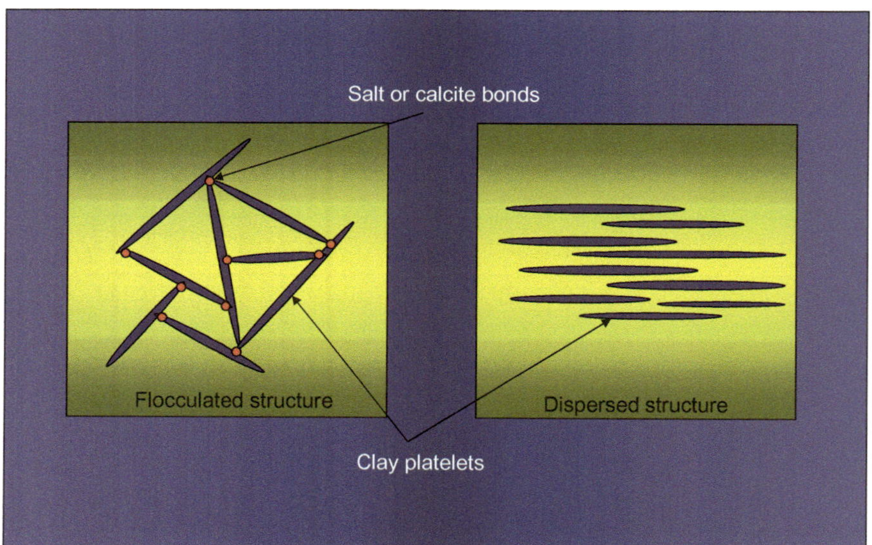

Figure 8.4 The figure displays typical clay structure. Once the salt or calcite bonds are dissolved by water, the structure collapses from flocculated to dispersed. It will never return to flocculated.

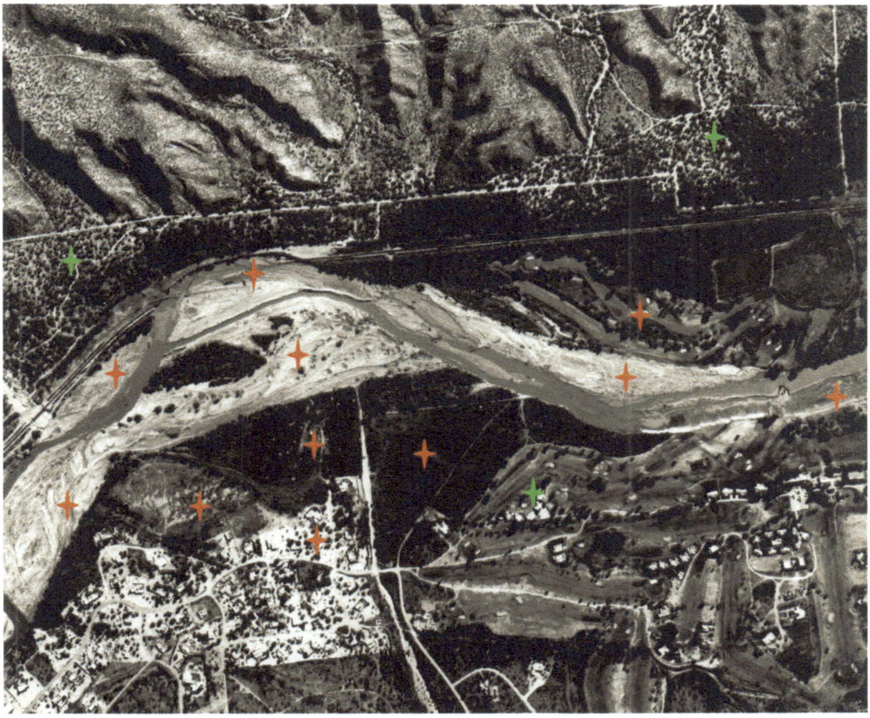

Figure 8.5 The vertical aerial photograph shows the Santa Cruz River near Tubac, Arizona soon after the epic 1983 flood.

In southern California, collapsing soils have been the cause of severe damage to aqueducts moving water from the Sacramento River delta area to Los Angeles.

8.3 GROUND SUBSIDENCE DUE TO "GROWTH FAULTS"

Growth faults are features common to large deltaic formations. In the United States, growth faults are most common within the Mississippi River deltaic sediment in Texas and Louisiana. Growth faults are non-tectonic and are actually large landslides occurring as unsupported deltaic sediment slumps seaward producing a head scarp on land as shown in Figure 8.6.

Although growth faults are not tectonic in origin, and do not produce earthquakes or tsunamis, there can still be significant damage to structures built atop them. As there is no feasible way to stabilize these

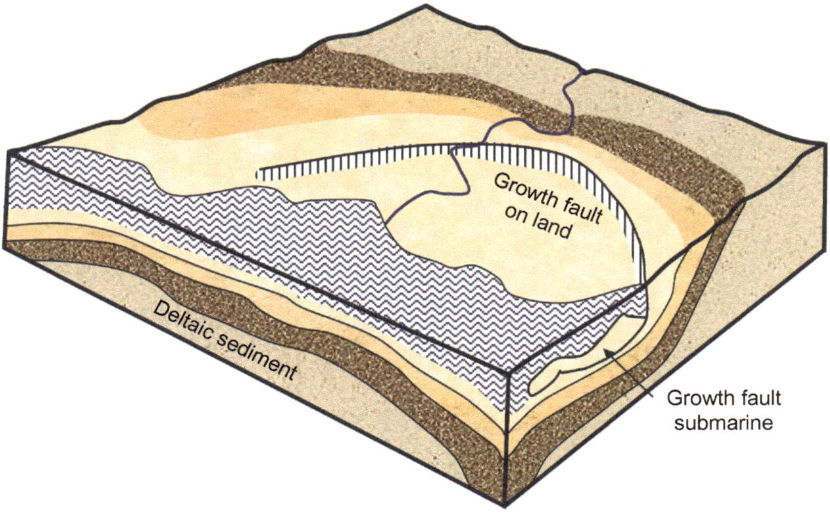

Figure 8.6 The cartoon depicts growth fault in deltaic sediment.

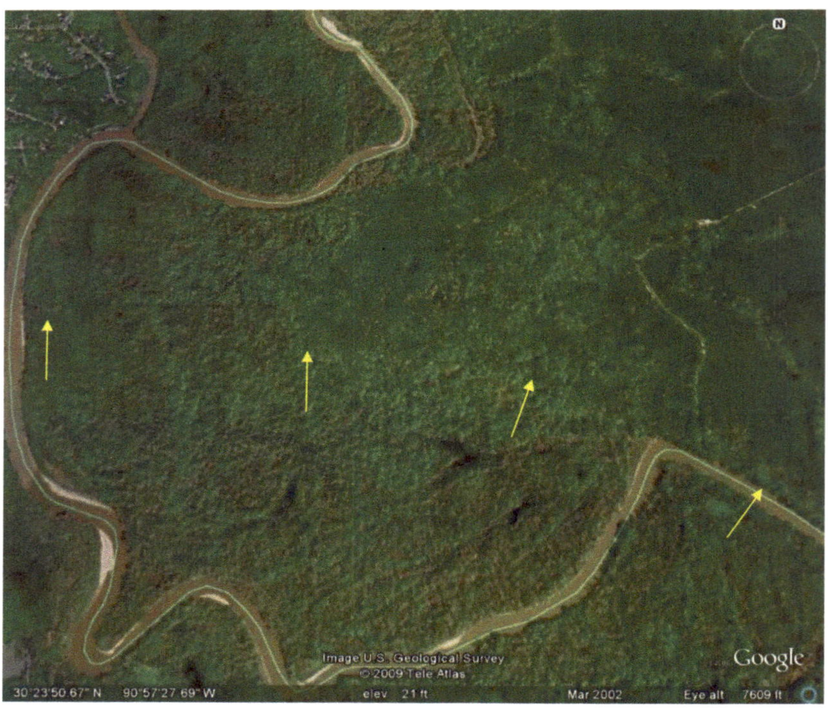

Figure 8.7 The vertical composite photo from Google Earth shows a large growth fault south of Baton Rough, Louisiana (yellow arrows). Growth faults form the head scarp for a large landslide within deltaic sediment onshore and offshore.

landslides, they continue to grow until a stable geometry is reached farther down the delta. Growth faults are difficult to see on aerial photographs without relying on stereo pairs to enhance topographic displacement. Figure 8.7 shows a large growth fault south of Baton Rough, Louisiana.

Vegetation helps mask the growth fault, but with stereo photographs, the large scarp can be readily seen. In Figure 8.7, yellow arrows are used to delineate the growth fault which is the head scarp of a large landslide occurring both onshore and offshore as deltaic sediment slumps into the Gulf of Mexico.

REFERENCE

1. Casavant RR. Personal communication, 2010.

PART III

Spectral Ranges Beyond Visible

We saw in Part I: Establishing Baselines, that panchromatic and color aerial photographs are convenient to use because they correspond to the visible portion of the electromagnetic spectrum, where patterns, shades of gray, and colors are familiar to us. One downside to these products is that most earth materials are not very colorful in the visible portion of the spectrum. As it turns out, they aren't very colorful in any other portion of the spectrum either. Modern photographic materials are now available with sensitivities from 0.25 μm in the ultraviolet to 0.925 μm in the near infrared, and scanners are available over far greater spans. Should we use them? That depends on what you're looking for, whether the products are already archived somewhere, and how much money you have. As a general rule, I would advise, *before you go to the expense of acquiring images in exotic portions of the spectrum, have a very good understanding of (a) how your target will appear on images acquired in other portions of the spectrum, (b) how your target will contrast with its background there, (c) resolution and fidelity characteristics of the images, and (d) geometric and fidelity aspects of the scanner.*

I say this because things do not look the same there. Patterns, colors, and tones that we have stored in our memory since birth are no longer valid there. So interpretation of images acquired there is closer to a serial process vis a vis a parallel process so common to image interpretation using the visible. The interpretation process is serial because we must constantly remind ourselves of the reasons for the occurrence of a given pattern or color. This takes time and close attention. Remember that most of the information flowing into your visual cortex flows from upper levels of your brain, not from your optic nerve. So before you interpret patterns acquired in nonvisible spectral regions, you must teach your brain their meaning.

I believe that, for most of your work, aerial photographs using the visible spectrum provide the cheapest, most valuable, most consistent, most available, most intuitive, and broadest coverage available. Who could ask for more? All right, there may be specific projects that could benefit from imagery collected beyond that which is visible to the human eye. So in the space remaining let's take a brief look at characteristics of images and earth materials (targets) collected in portions of the electromagnetic spectrum beyond what we can see with our eyes.

The following chapters progress from the highest frequency (shortest wavelength) ultraviolet portion of the spectrum to the lowest frequency (longest wavelength) microwave portion of the spectrum. Unfortunately, we can't just run off and choose any portion of the spectrum for our imaging. There are some things about solar energy, earth energy, sensing systems, and the atmosphere that will determine where we can and cannot look. I have put together Figure III.1 to serve as an illustration.

Figure III.1 shows a model of the spectral component of sunlight hitting the earth assuming the sun is a 6000 K blackbody (not a bad assumption) and there is no atmosphere (not a good assumption). Note that the solar spectrum (pink color) peaks at approximately 0.55 μm in the green portion of the spectrum and falls off rapidly thereafter. Also shown on the graph is an approximation of the sunlight actually striking the earth after passing through our atmosphere (blue color). This is important, because the atmosphere is not friendly to all electromagnetic radiation. Within the near infrared, for example, water and carbon dioxide in the atmosphere absorb sunlight at several wavelengths. Within the ultraviolet, ozone at the upper reaches of the atmosphere completely blocks all sunlight having wavelengths shorter than approximately 0.28 μm. That's good for our skin, but a real limitation for remote sensing in the ultraviolet.

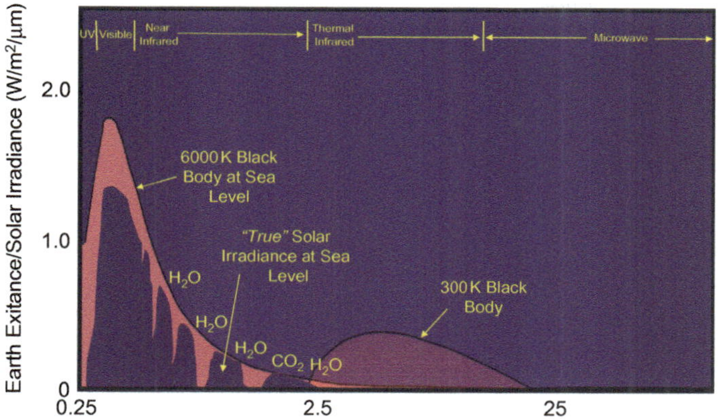

Figure III.1 The composite figure displays the modeled solar irradiance of a 6000 K blackbody (temperature of the sun) at the top of the earth's atmosphere, the actual *solar irradiance at the earth's surface after atmospheric absorption, and a model of the earth's exitance (energy produced by a 300 K blackbody), also at the earth's surface. I have increased the relative amplitude of the earth's exitance so it can be seen.*

At longer wavelengths, the heat emitted by the earth (300 K blackbody) can be seen peaking at approximately 9.6 μm. This area is within the thermal infrared portion of the spectrum. Beyond these energy sources, in the microwave, there is little ambient energy available there, so we must generate our own.

Photographic products are available for use in the ultraviolet, visible, and into the near infrared. Once we get beyond the near infrared and into the thermal infrared, however, we are beyond the realm of film emulsions, and must use scanners or fixed arrays of sensors sensitive to the low energy available there.

CHAPTER 9

The Ultraviolet

The ultraviolet comprises high frequency radiation adjacent to the blue hues of the visible portion of the electromagnetic spectrum (Figure 9.1).

Ultraviolet energy was first discovered in 1801 when Johann Wilhelm Ritter discovered that silver chloride (an important component of panchromatic film emulsions) actually degrades faster in the dark portion of the spectrum adjacent to the violet than anywhere in the visible. The ultraviolet is commonly divided into the regions shown in Table 9.1.

Two main factors limit our use of the ultraviolet for hazard recognition. First, atmospheric scattering is severe in the ultraviolet, making photographs (incidentally, panchromatic films can be used in the ultraviolet) fuzzy and of low contrast. Second, rock reflectance plunges in the ultraviolet, so that rocks that may be bright in the visible are dim in the ultraviolet, reducing still further the possible contrasts on aerial photographs. To a large extent, then, expect images in the ultraviolet to be fuzzy and of low contrast.

The reflectance of some rock types, such as carbonates, drops less precipitously in the ultraviolet than most other rocks providing reasonably good contrasts, but I would be surprised if the two disadvantages mentioned above don't overwhelm this minor advantage.

Another advantage of ultraviolet images relates to the transmission properties of water. Unpolluted water transmits ultraviolet energy much better than energy at longer wavelengths. Hence, ultraviolet energy propagates farther through water than does energy in any other region of the spectrum. That is why you can receive a sunburn even though you are under water, and why you need filters for proper color rendition of underwater photography. If you are interested in details of shallow offshore features, the ultraviolet may be the region for you.

There has also been some research that shows the ultraviolet to be a promising region for detecting vegetation stress and pollution in aquatic systems. This research uses a device called a Fraunhoffer line

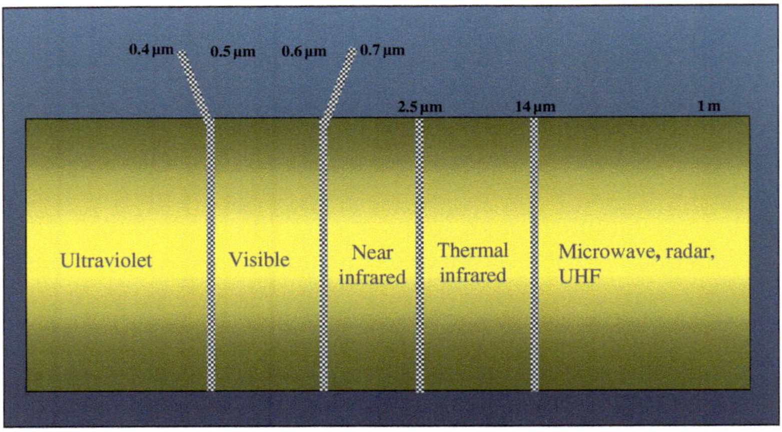

Figure 9.1 The figure presents a portion of the electromagnetic spectrum.

Table 9.1 Ultraviolet Regions

Designation	Wavelength Range (μm)
UVA	0.32–0.4
Near	0.3–0.4
UVB	0.28–0.32
Middle	0.2–0.3
UVC	0.1–0.28
Far	0.122–0.2
Vacuum	0.01–0.2
Extreme	0.01–0.121

depth discriminator. Fraunhoffer lines are narrow areas of the spectrum within which gasses around the sun have absorbed all energy. Any energy within a Fraunhofer line must originate from the earth. This can occur when energy in wavelengths adjacent to the Fraunhoffer line is absorbed by earth materials and reemitted at longer wavelengths within the Fraunhoffer line. The depth of the Fraunhoffer line is a measure of the amount of energy measured there, the deeper the line, the lower the energy. Fraunhoffer line depths have been seen to be sensitive to pollution in water bodies, and when calibrated using known substances, are capable of distinguishing concentrations to a few parts per million. This technique may also be more sensitive to vegetation stress than near infrared images.

Although underutilized at the present, the ultraviolet could one day prove to be a tool for natural resource managers (conservation and extraction arenas) as well as for managing natural resources such as forest ecosystems, wildfire planning, agriculture and land use, drought management, and contaminant inventory and monitoring.

CHAPTER 10

The Near Infrared

The infrared comprises low frequency radiation adjacent to the red hues of the visible. We usually divide the infrared into two regions, the near infrared (NIR) and the thermal infrared (TIR) as shown in Figure 10.1.

The near infrared was discovered in 1800 by Sir William Herschel. Herschel was interested in the temperature of different colors. He placed a thermometer in the colors produced by a prism and found that the hottest region occurs in the dark portion of the spectrum beyond the red. He termed this "hot light," and it was later labeled the infrared. As Herschel documented, the infrared is adjacent to the red portion of the visible spectrum and photographic products are available there. For this reason, the NIR is often referred to as the *photographic IR*. The brightness we see in the NIR is due to reflected sunlight, just as it is in the visible. Despite what you may glean from Herschel's temperature measurements, the brightness of NIR images is *not due to the temperature of a target!* It is due to the reflectance of that surface to NIR radiation. Never make such a mistake, even though many people do so. Since brightness in the NIR is due to reflected sunlight, images of rocks and soil in the NIR look similar to those in the visible. If you look back to the spectral reflectance plots of Chapter 4 (Figures 4.1, 4.2, and 4.8), you will note that the reflectances increase and separate slightly within the NIR (beyond 0.7 μm). Hence, there may be slightly more contrast among rocks in the NIR—but not much. This is **not** the reason to choose NIR photography. Instead, I invite you to take a look at Figure 10.2 to see why the NIR may be of interest to you.

Figure 10.2 shows a redrawing of Figure 4.1. In Figure 10.2, the spectral reflectance of the green shirt is replaced by the spectral reflectance of vegetation. A couple of things are important here. First, note that the reflectance of vegetation is low in the ultraviolet and blue portions of the electromagnetic spectrum. This is due to strong absorption at these wavelengths by leaf pigments, chlorophylls, and carotenes. Further, in the

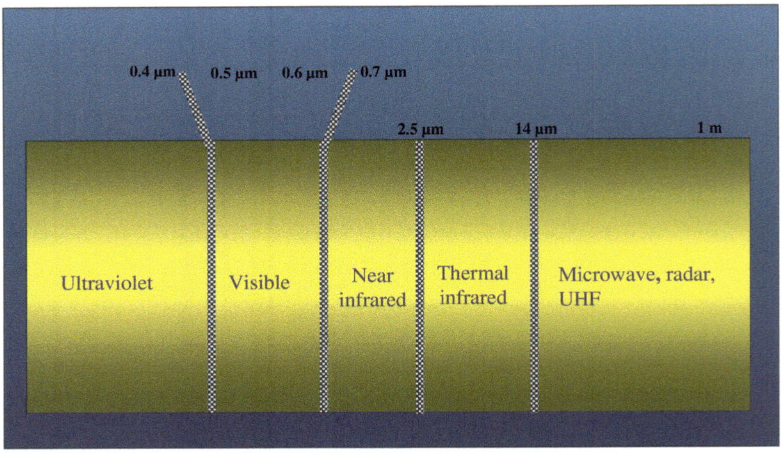

Figure 10.1 The figure shows a portion of the electromagnetic spectrum. As you can see, the range of wavelengths for the different regions has been reduced considerably to fit in the figure.

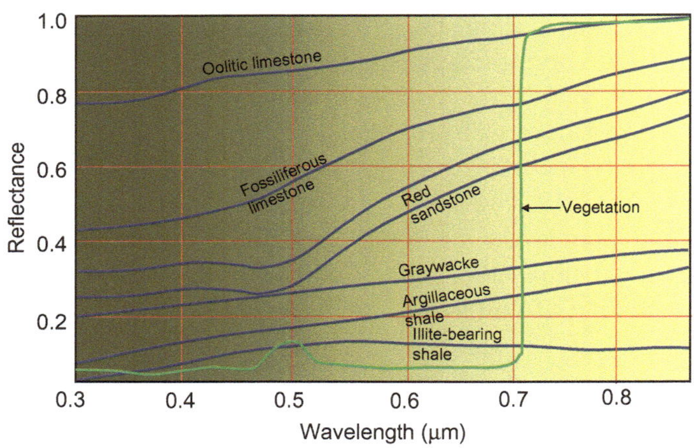

Figure 10.2 This is a redrawing of Figure 4.1 showing the spectral reflectance of sedimentary rocks with special emphasis on the spectral reflectance on vegetation rather than a green shirt.

green the same chlorophylls and carotenes produce a small increase in reflectance. This, of course, is why vegetation looks green to the human eye. The increase is not very great, though, if you think about it you will agree that vegetation is not very bright; you never have to squint, for example, when you look at a plant. Reflectance drops again as the chlorophylls and carotenes absorb strongly in the red.

Finally, look what happens in the NIR *just* beyond the sensitivity of the human eye at 0.7 μm. The reflectance of vegetation increases abruptly. This behavior has been termed the "*red edge*" as it is steep and occurs just beyond the red portion of the spectrum. The abrupt increase in reflectance in the NIR is not due to chlorophylls and carotenes, but rather to much more intrinsic properties of the plant being imaged. The high vegetation reflectance in the NIR is a function of such things as leaf type, leaf structure, leaf stress or general *quality of life*, leaf water content, leaf pubescence, and other intimate things known only to the leaf. The slope of this red edge tells us a great deal about the type of plant being imaged as well as its general health. As fall approaches, for example, the red edge of a deciduous plant begins to flatten, extending into the red somewhat. So now we can see a little of the red edge in the red portion of the spectrum between 0.6 and 0.7 μm. When our eye combines the little bit of red with the green due to chlorophyll, we get yellow. The fall colors that photographers and outdoor enthusiasts love so much are due entirely to the flattening of the red edge as leaves die, and its subsequent appearance within the visible portion of the spectrum.

So the NIR is valuable for its sensitivity to vegetation type, density, water content, and general health. This is an especially useful region because film emulsions are available for imaging in the NIR, bringing with them the same photographic fidelity and parallax so valuable in the visible portion of the spectrum. There is one minor problem, though. When we make color film emulsions for visible region photography, we use three layers of emulsions, one sensitive to red wavelengths, one sensitive to green wavelengths, and one sensitive to blue wavelengths. All other colors can be made by combinations of red, green, and blue (a *tristimulus* color coordinate system).

When we make color infrared film emulsions for NIR photography, we need four layers of emulsions, one sensitive to NIR, one sensitive to red, and one sensitive to green. The fourth layer is necessary because the NIR emulsion is also sensitive to blue, so we need a layer to absorb blue wavelengths before they hit and interact with the NIR layer. Here is where the minor problem occurs. To what colors within the tristimulus color coordinate system do we assign the three active layers? Color is traditionally assigned as follows:

- NIR ← red
- Red ← green

- Green ← blue
- Blue ← filtered out

Do you now see what I meant earlier when I stated that things with which you are familiar look different using nonvisible portions of the spectrum? For this reason you must continuously think about the meanings of the colors you are trying to interpret. It is always a good idea to give some thought to what your target is and how you expect your target to appear in the NIR, *before* you order the images. See Figure 10.3 to get an appreciation for the way things appear in NIR images.

The left image in Figure 10.3 is a standard color vertical aerial photograph taken with standard color film in the visible portion of the spectrum. The photograph on the right is a color vertical aerial photograph taken with color infrared film in the NIR portion of the spectrum. Note the brightness of the NIR photograph and compare the reflectance of vegetation in the NIR with the much lower reflectance in the visible (Figure 10.2). The pair of photographs makes for a great comparison, but the photograph on the right has been altered. Apparently someone has digitally stretched the contrast of the photograph (this can be easily done in a program such as Adobe Photoshop™), because the silt-laden stream should be blue, not black, in the NIR color photograph. Remember that the color blue is

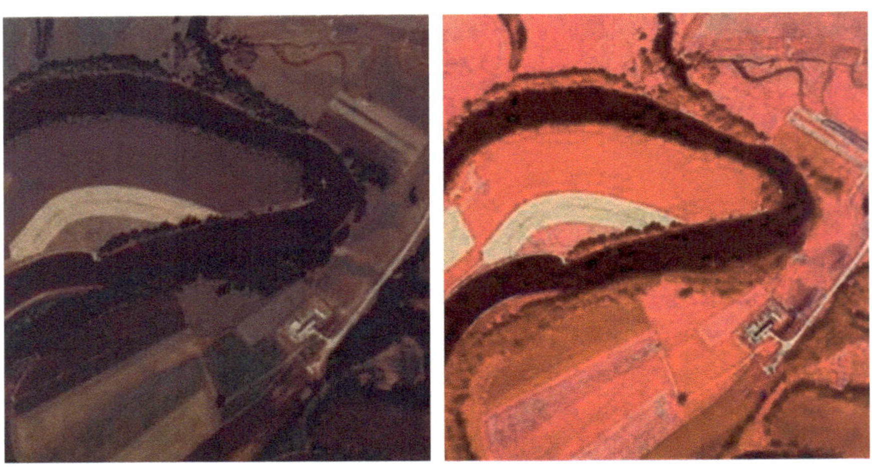

Figure 10.3 These vertical aerial photographs display a comparison between a color visible photograph (left) and a color NIR photograph (right). The photograph on the right has been altered to make the stream appear black. The silt-laden stream should appear blue. Both photographs appear on www.geomart.com as examples of their sales inventory.

traditionally assigned to green. The greenish brown of the silty stream would, under unaltered conditions, appear blue, e.g., see Figure 10.4.

There are a few additional advantages to photography in the NIR. First, the longer wavelengths used for color NIR photography coupled with the filtering of blue wavelengths, avoid most of the atmospheric

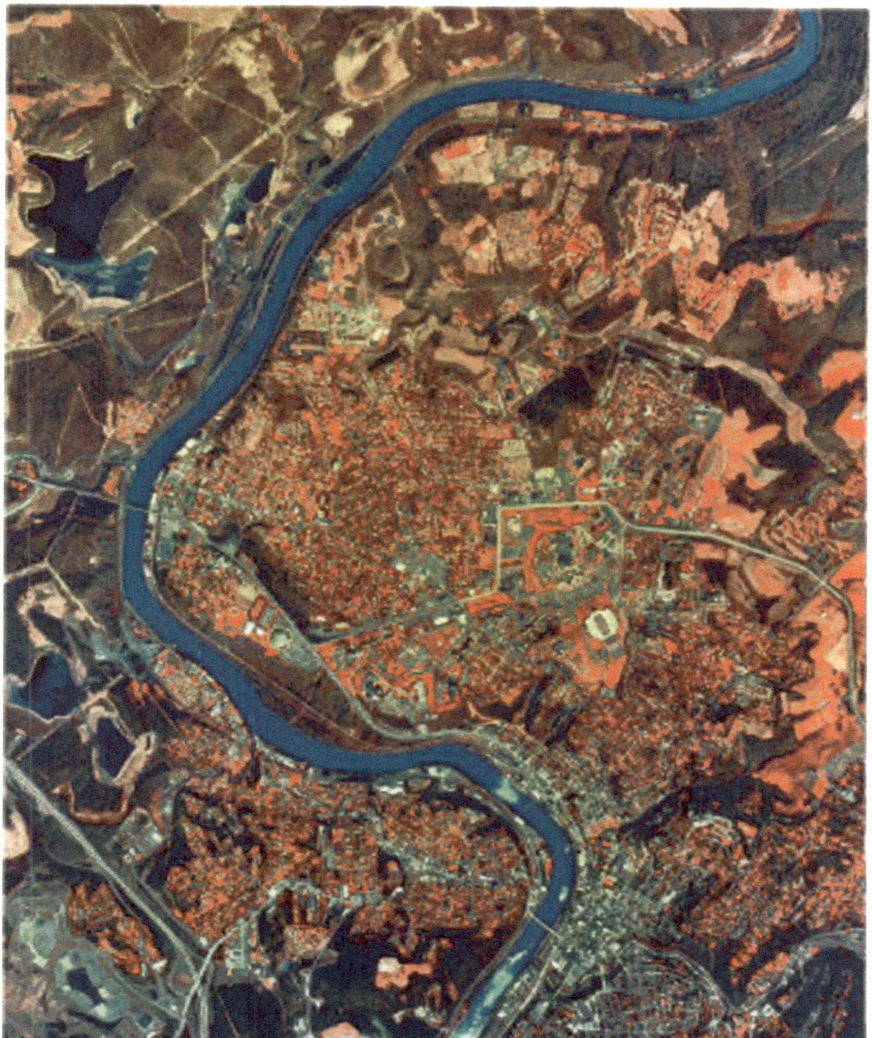

Figure 10.4 This is another example of NIR color photography from www.geomart.com. This photograph has not been stretched, as you can see the blue of the river due to silt.

scattering common to the short wavelength blue portion of the spectrum resulting in sharper, crisper photographs. Second, although one can design camouflage coverings using high NIR reflectance dyes or paints, camouflage coverings with normal dyes would be obvious in NIR photographs. Third, the transmittance of water to NIR wavelengths is very low. So, ultraviolet energy penetrates water to impressive depths (meters), but NIR energy does not (less than a centimeter). If your aim is to accurately measure water–land interfaces, for example, to verify the extent of flood inundation for insurance purposes, the NIR is a good choice.

Along with advantages there are usually some disadvantages. You will pay more for color NIR photographs. Camera exposure properties are a little trickier in the NIR than in the visible, so you will want to verify that the camera operator is experienced with NIR photography. Archives of existing NIR photographs are sparsely populated outside of government laboratories (see the websites in Chapter 13).

CHAPTER 11

The Thermal Infrared

The thermal infrared (TIR) extends from approximately 2.5 to 14 μm (Figure 11.1).

Atmospheric absorption dominates the TIR with the exception of a window from approximately 2.5 to 5 μm, and from 8 to 14 μm. The 2.5–5 μm window contains approximately equal portions of reflected sunlight and earth radiometric temperature (see Figure III.1, page 135). This combination makes attempts to correlate radiometric temperature (amplitude or brightness on an image) with thermometric temperature (temperature measured by a thermometer) difficult. The 8–14 μm window corresponds to the peak of the earth's exitance with little solar interference, so such temperature correlations are more reasonable in this window.

An important thing to keep in mind is that there ***is not*** a great deal of signal amplitude in this region. Certainly things heat up during the day, but that is primarily due to shadows, cool vegetation, and solar energy being preferentially absorbed by dark objects, later to be reemitted at thermal wavelengths (you may wish to review the discussion on reflectance, absorptance, and transmittance in Chapter 1). Thermal images generally have higher contrast during the day than at night, but if all we are going to do is record shadows and differences in absorptance, and by corollary reflectance, then why not simply use color film which is far superior and cheaper than thermal images? I will provide a partial answer to this question later, but in all but that one situation, I will stick to the following advice—always employ thermal imagery only at night or, better yet, at pre-dawn before the sun begins to heat up the targets. Pre-dawn imagery lets the pilot of the scanning aircraft better see where he is going, yet maintains nighttime thermal properties.

So what's the problem with nighttime or pre-dawn imaging? The problem is that there is not much thermal energy out there. Think of it this way—There are few thermal photons flying around out there at night. So given this fact, how can we best collect them? One way is to get a very large collector (this is termed as a large aperture system), and collect thermal photons from a large area of the ground. This works

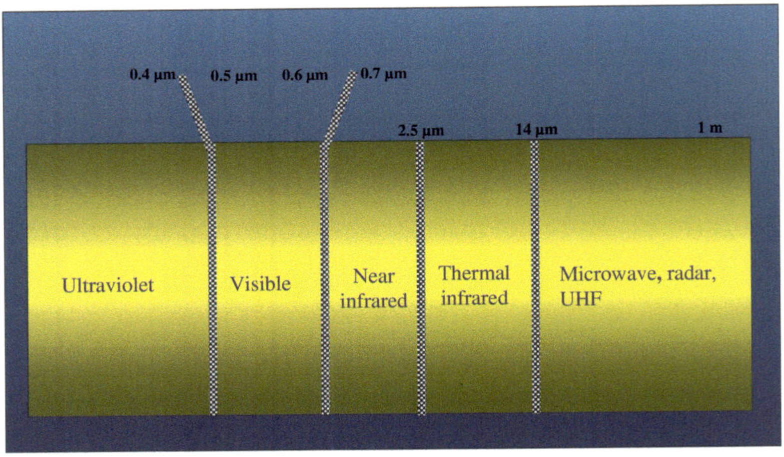

Figure 11.1 The figure presents a portion of the electromagnetic spectrum.

fine if you wish to maximize your signal resolution, but in doing so you sacrifice spatial resolution (you may wish to refer back to the discussion of resolution in Chapter 1). Or, you may wish to use a small collector (small aperture system) to improve your spatial resolution, but in this case you sacrifice signal resolution (you're not quite sure how the brightness of different targets relate to each other quantitatively, because there are so few photons collected). You can not have both unless you have the time to let the small collector sit in one place and collect photons for a long time. This is a viable strategy for astronomical observations with telescopes, but it is not very practical for airborne systems. So we have to decide whether we want high spatial resolution or high signal resolution.

11.1 SCANNERS

Why do we need scanners? Why not just use film as we did in the visible and NIR. Two reasons, first we could probably coat a piece of acetate with an emulsion sensitive to heat, but the mechanical operation of the camera would surely generate enough heat to fog the film, rendering it useless. Second, cooling a large film surface is prohibitively expensive, so we need some type of scanner for collecting images in the TIR. Most of the early scanners were composed of a heat sensitive material about the size of the head of a pin at the end of a wire. Such an apparatus can be inexpensively cooled. Photons of TIR energy

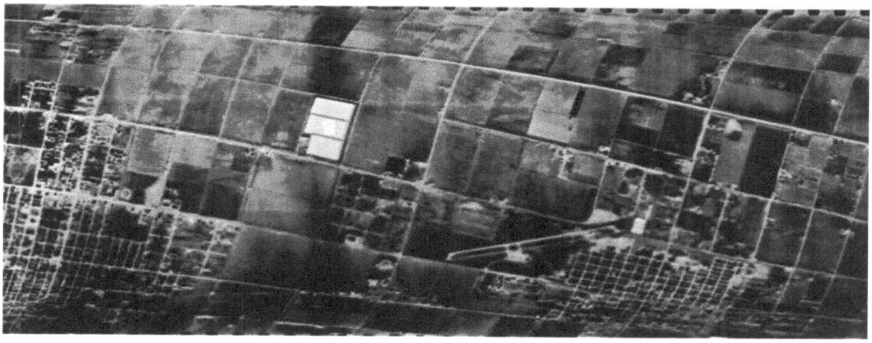

Figure 11.2 The figure from HRB-Singer, Inc. is a daytime whisk broom thermal scan of Imperial Valley, California. Note the "S"-shaped mapping of the orthogonal road system.

would strike the heat sensitive material and induce a current to flow in the wire. The amount of current generated was proportional to the number of photons striking the sensor. The current was measured as a number and stored forming a digital image. The numbers could later be used to modulate a light beam to expose standard film for interpretation.

Whisk broom scanners—Whisk broom scanners act like a whisk broom, sweeping side to side as the aircraft flies forward. Many of the early archived TIR images were collected using whisk broom scanners. There are some serious problems with these scanners. First, the scanner must slow down and stop at the end of a sideways sweep before it can sweep back the other way. The maximum scanner speed is in the middle of the sweep. The resulting image is distorted by this activity producing a geometry that looks like a lazy "S." If you think about it, the scanner records a temperature at the far left edge of the flight line, then as it sweeps right, the plane moves forward and a new area is imaged. At the end of the sweep the scanner collects temperature at the far right, but by now the aircraft has moved quite a bit along the flight line (Figure 11.2).

The light tones in Figure 11.2 are warm relative to the dark tones. Roads are warm, vegetated fields are cool. If details of the flight speed and scanner speed are known the thermal image can theoretically be digitally corrected, however, most attempts fail. Note I indicated earlier that light tones are warm relative to dark tones. I made no attempt to assign a temperature to the different tones. There is a good reason for this.

Calibrating thermal sensors is exceedingly difficult. The best approach for calibrating thermal images involves a grid of thermometric measurements on the ground at the same time the imaging mission is flown.

Push broom scanners—Most new scanners are of this type. Basically, a linear array of sensors is pushed along the aircraft flight path, or a matrix of sensors is used to acquire instantaneous 2-D images. These scanners produce less distortion than whisk broom scanners, but both scanner types are subject to distortions due to aircraft pitch and yaw and both share the same calibration problems and relatively high expense.

I promised to reveal one good reason to use thermal scans during the day. Actually it is the only reason I can think of. A look at the cool vegetation signatures in Figure 11.2 will give you a hint. If it is only vegetation that you are looking for, the NIR is a better region for your sensing, but if you are looking for areas having near-surface moisture, especially without vegetation, thermal images may provide good information. Faults and fractures often form barriers or conduits for near-surface moisture. Insipient landslides also occasionally disrupt near-surface moisture. If you believe that this is possible within your study area and you can't see the faults, fractures, and landslides by any other imaging technique, you might consider giving a thought to TIR.

One of the best sources of archival thermal images is probably data from AVIRIS (Advanced Visible InfraRed Imaging Spectrometer). AVIRIS is an airborne whisk broom scanner, with images confined to the near-nadir portion (middle) of the scan to limit distortion. There is a limited amount of archival images and some free images are available at http://aviris.jpl.nasa.gov. Archival images were not available for order as of June 2009 but may be available soon. New AVIRIS projects require a proposal to NASA. Further details are available on the website.

11.2 SUMMARY

Here are some things to think about before ordering a thermal scan.

- Thermal scans are expensive, and contractors willing to conduct them are few. So make sure that there is a very good reason to expect that thermal contrasts will give you an advantage in recognizing hazards on the ground.

CHAPTER 12

The Microwave

The microwave portion of the spectrum contains the longest wavelength energy that will be used in most remote sensing projects (Figure 12.1).

The most common remote sensing in the microwave portion of the spectrum is done using radar (Radio Detection And Ranging). Natural sources of electromagnetic radiation in the microwave region are of such low amplitude that there is little natural noise or interference in this region, yet we are able to construct systems capable of generating microwave radiation at high intensities. In addition, the atmosphere (even when cloudy) is transparent to the long (1 cm to 1 m) wavelengths used in radar imaging, so for all practical purposes radar is an all-weather sensing system.

Radar imaging is accomplished by mounting a radar antenna on the wing of an aircraft and emitting radar pulses out to the side. For this reason, the common term used for such a system is *S*ide *L*ooking *A*irborne *R*adar or SLAR. An SLAR image may be formed by modulating a light beam by the amount of radar energy reflected back to the aircraft from the ground surface. This light beam exposes common film as the aircraft flies along its flight path.

Although an SLAR image looks very much like a vertical aerial panchromatic photograph (Figure 12.2), don't be fooled; light and dark tones do not mean the same thing.

The most striking dark tones on the image correspond to the rivers. Why so? Remember two things. First, the long wavelengths in the microwave region do not penetrate water well. Second, the radar image is side looking, so that the radar pulse strikes the water and reflects specularly (like a mirror) away from the aircraft, never to return to be collected. Figure 12.3 will make this easier to see.

So how does Figure 12.3 relate to remote sensing in general and to radar specifically? The words "relative to λ" in Figure 12.3 are important to our discussion. Most surfaces, for example, are rough when

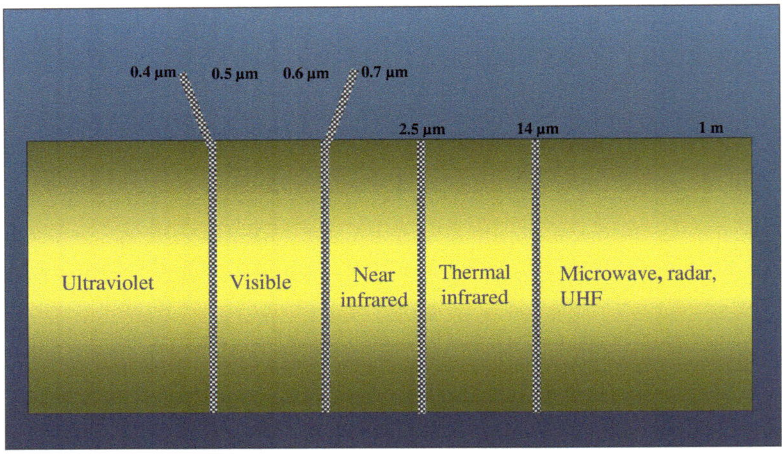

Figure 12.1 The figure presents a portion of the electromagnetic spectrum.

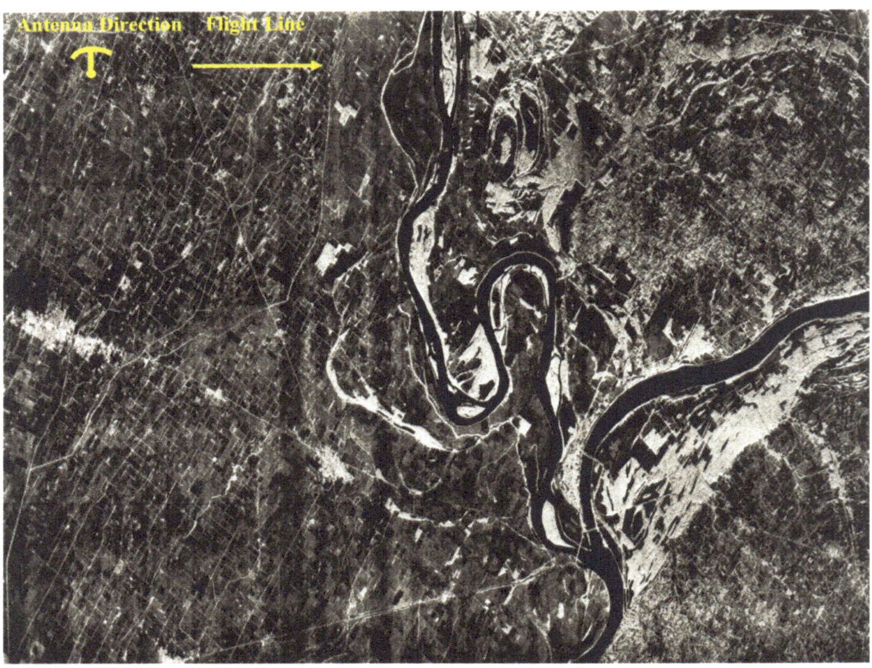

Figure 12.2 The figure shows a side-looking airborne radar image of the Cairo, Illinois area at the junction of the Mississippi and Ohio rivers. North is to the top of the figure, the antenna points south and progresses from west to east.

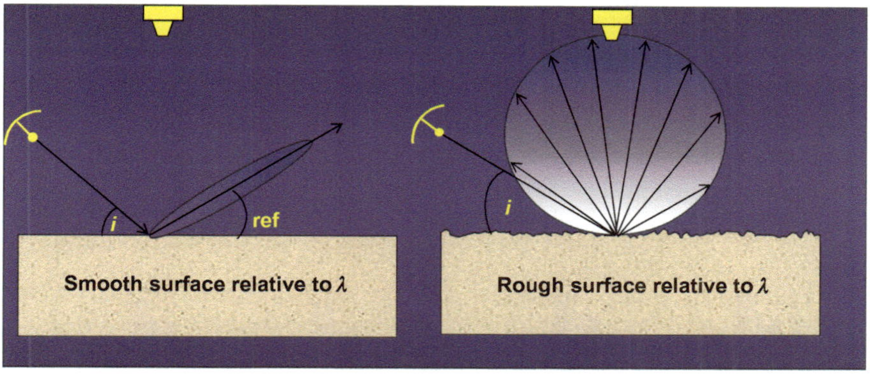

Figure 12.3 The figure shows the ways in which a surface reflects electromagnetic energy. A smooth surface (relative to the incident wavelength, λ), reflects energy as a specular (mirror-like) reflection with the angle of reflection (ref) equal to the angle of incidence (i). A rough surface scatters reflected energy at many angles. The radar antenna is shown at left, and a typical camera angle at top.

considering visible wavelengths (0.4–0.7 μm), so reflected sunlight is scattered in all directions by most surfaces (calm water, and mirrors are exceptions). A surface that is rough when considering sunlight, however, may be smooth to radar wavelengths (on the order of many centimeters), thus sending the radar energy off away from the antenna.

Remember also that we are dealing with a *side-looking* imaging system with radar. When imaging in the visible and NIR, the energy source (antenna on the left of Figure 12.3) is the sun, and the image is collected using a receiver (camera) situated over the target surface (shown at top). With radar the source antenna also serves as the receiver, hence, a smooth surface to radar will reflect in a specular manner and not be returned to the antenna (no return = black). Only a rough surface will return energy to the source antenna. So what else determines the radar image brightness?

Figure 12.2 is actually a collection of image strips collected along east-west flight lines. In each flight line, the radar antenna is pointed south as shown at the top of the figure. Variations in brightness on radar images are due to the following factors:

- *Topography*—The side-looking aspect of radar can not see the back sides of steep hills, so there is no return (black image) from these surfaces. Topography usually provides most of the contrast on a radar image.

- *Surface roughness*—On surfaces not hidden by topography, surface roughness, as discussed earlier, provides most of the variations in brightness on a radar image. Whether a surface is rough or not depends on the following:
  - Surface irregularities
  - Radar wavelength
  - Radar incident angle
  - Radar polarization.

  Vegetation usually provides the primary contribution to surface roughness.
- *Radar wavelength*—The wavelength of the radar helps determine if a surface will be perceived as rough or smooth.
- *Surface electrical properties*—Metals provide the strongest radar reflections, but healthy vegetation is also a strong reflector.
- *Radar incident angle*—Objects that are orientated (walls and corners are the most common) in such a way that they reflect most of the energy straight back to the antenna provide strong return signals on radar images.
- *Radar polarization*—Antennas can be set up to transmit and receive both horizontally and vertically polarized signals.

So let us look at some of the features on Figure 11.2. We have already decided that the dark tones of the Mississippi and Ohio rivers are due to specular reflection of the radar away from the antenna. The strong returns adjacent to the rivers are due to vegetation. The strong returns in the small communities along the highways also provide bright islands within the radar image. How would you explain the brightness of the rural road system?

Here are some other things to think about before purchasing radar.

- Radar is an all-weather sensor. It is therefore extremely valuable for areas where cloud cover is a problem throughout the year.
- In forested and heavily vegetated areas, the radar radiation reflects from the tops of trees and bushes. Because of the resolution you may think you are looking into the ground when you are actually looking at the top of the vegetation canopy. On the positive side, the vegetation canopy usually drapes topography, and topography is usually a good indicator of geologic structure. To the extent that the "*usually*" is true, radar can provide good insight into geologic structure.

- A good rule of thumb is that radar images have a spatial resolution of from 10 to 15 m. Those involved in using research systems at low altitudes may claim higher resolution, but I would not trust archival products beyond 10–15 m.
- As radar images are strongly influenced by topography, you may consider whether digital topographic data may provide you with the same information at a lower cost.
- The process for generating digital topography has been greatly enhanced recently by the development of airborne LIDAR (Light Detection And Ranging) topographic mapping systems. LIDAR uses shorter wavelengths than RADAR (UV through visible to TIR), thereby rendering digital topography at higher resolutions. LIDAR can also be used in a backscatter (imaging) mode as compared with a ranging (topographic) mode to produce 3-D color images. One should not underestimate the amount of computer processing necessary to render LIDAR 3-D images, however, so at present their cost is high and their availability is low. Still, there *is* a promising future for LIDAR.

CHAPTER 13

Obtaining Aerial Photographs

There are a number of places to obtain archived aerial photographs and order aerial photographic missions. I have listed several of the most convenient below.

13.1 US DEPARTMENT OF AGRICULTURE

Previously there were several agencies within USDA that offered aerial photographic services. These have been consolidated into a single agency, which offers both archived aerial photographs and remote sensing images. One can also request new photo missions. http://www.fsa.usda.gov/FSA/apfoapp?area = home&subject = prod&topic = cat.

13.2 US DEPARTMENT OF COMMERCE

Aerial photographs within the US Department of Commerce are handled by the National Oceanic and Atmospheric Administration at http://geodesy.noaa.gov/PC_PROD/catalog/aerial_photos.htm.

13.3 NATIONAL ARCHIVES AND RECORD SERVICE

http://www.archives.gov/research/order/maps.html.

13.4 US DEPARTMENT OF THE INTERIOR

US Department of Interior remote sensing products, both archived and new projects, are handled by the US Geological Survey at http://www.usgs.gov/pubprod/.

13.5 CANADIAN AERIAL PHOTOGRAPHY

http://airphotos.nrcan.gc.ca/photos_e.php Click on NAPL online.

13.6 MEXICAN AERIAL PHOTOGRAPHY

I could find no clearing house for aerial photographs of Mexico. Your best bet may be to check with mining companies, or local firms in Mexico. Also, Google Earth, and Landsat images may be helpful.

13.7 FOREIGN AERIAL PHOTOGRAPHY

Many foreign countries are sensitive about anyone besides their military collecting aerial photographs of their country. I have seen foreign aerial photographs on which "*sensitive*" installations were redacted on some flight lines, but the censors forgot to redact the same installations on parallel flight lines. My guidance would be not to give up, but to expect significant bureaucratic interference from most foreign countries. Again, Google Earth and Landsat may provide valuable resources.

13.8 GOOGLE™ EARTH

A great resource for aerial images is Google™ Earth. Google™ Earth software can be downloaded free and provides images distributed worldwide. The images are satellite images supplemented with aerial photograph mosaics where available. I have used Google™ Earth images for projects in Papua, Indonesia, and Mexico where aerial photographs are difficult to procure. You don't get stereo images, but you can get 3-D perspective views of buildings and terrain using the 3-D buildings layer on the viewer. Images are in color. I have included Google™ Earth images in this e-book as Figures 5.1, 7.3, and 8.7.

13.9 OTHER SOURCES

1. Municipal, county, and state geological surveys and websites.
2. County, state, and national park system resource archives.
3. Municipal, county, and state water resources and flood management surveys and websites.
4. Municipal, county, and state departments of transportation.
5. College and university websites.
6. Local aerial photography companies.
7. Local and regional mining, petroleum, pipeline, railroad, and construction companies. Mineral extraction and pipeline companies

are constantly updating their photographic coverage to provide current topographic maps of their operations. Most of the aerial photographs from these projects are available from the contractor who flies the missions, but you will need permission from the owner of the photographs (e.g., mining company) to acquire them. Also, check the photo archives of the larger companies who consult for these commercial resource companies. They may prove fruitful from time to time.

APPENDIX

A Selection of Aerial Photographs for You to Interpret

In this appendix, I have placed six extra stereo pairs for you to enjoy. For each pair I have indicated the reason I selected it, but I will leave the interpretation to you.

Vulcan's Throne, Arizona. See if you can identify the potential hazards on this stereo pair. What kind of fault (normal, reverse, strike slip) passes through the cinder cone?

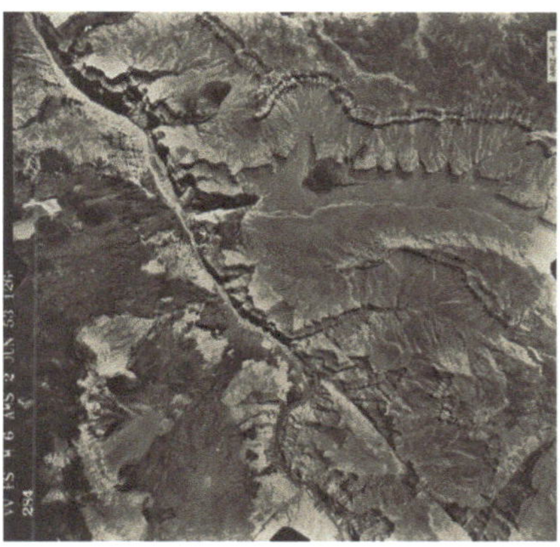

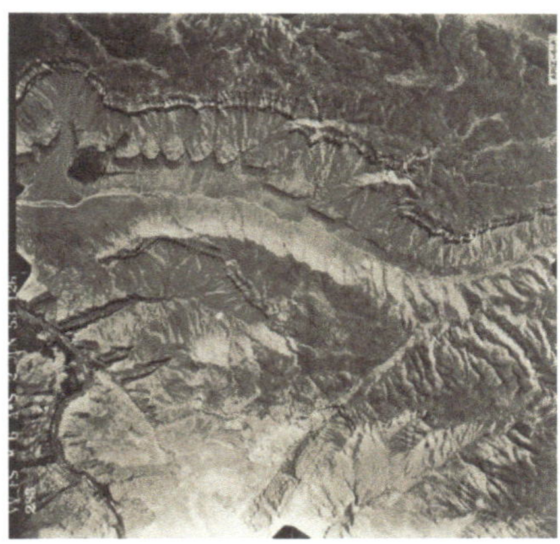

Death Valley, California. This stereo pair of the Hanaupah Canyon alluvial fan at the base of the Panamint Range indicates several potential hazards. Can you find them? I believe the feature at the upper left corner of the photos is a debris flow. My students insisted it is merely an old fan surface. How would you determine the correct answer on the ground?

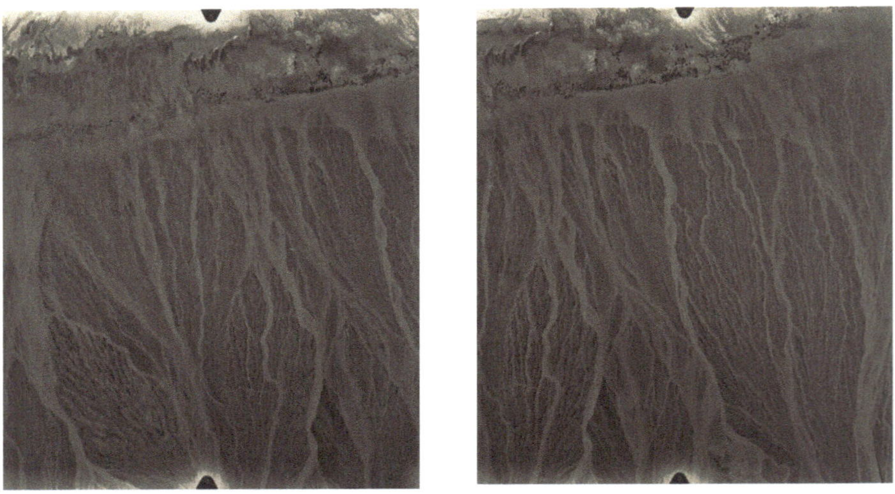

Columbia River, Washington. This stereo pair is of the Columbia River near Wenatchee, Washington. Recall the discussion about catastrophic floods in Chapter 7. Can you find evidence in these photographs to support the catastrophic flood hypothesis?

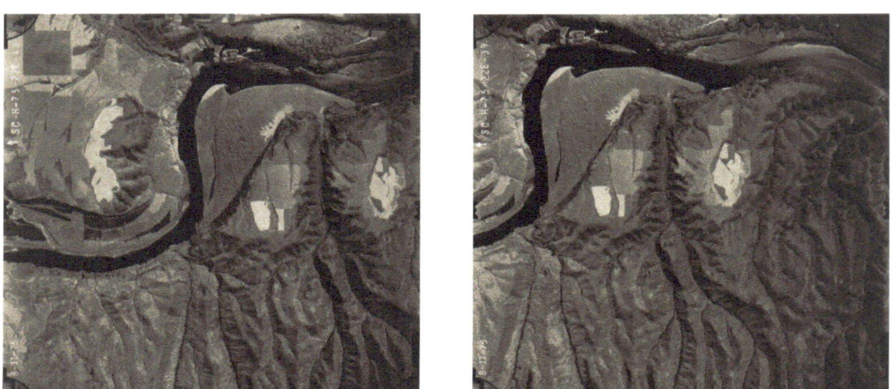

Columbia Plateau, Washington. This stereo pair near Yakima, Washington is part of the Columbia Plateau. What potential hazards do you see on the stereo pairs? Try to develop relative age relationships.

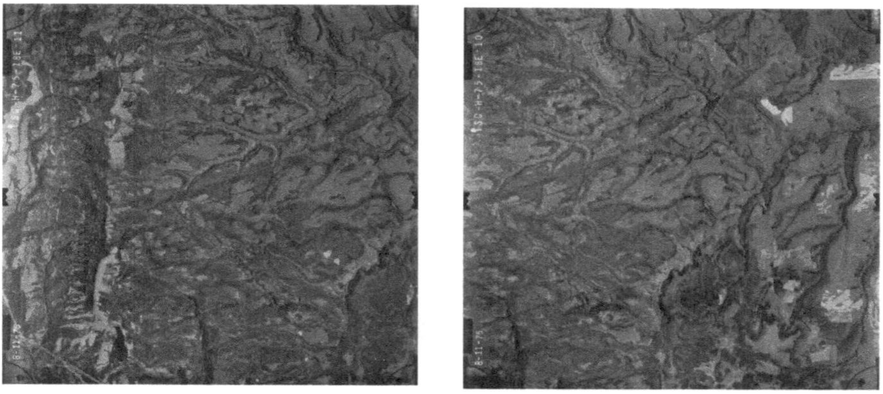

Cle Elum, Washington. This stereo pair is of an area near Cle Elum, Washington. A major mall is planned for construction at location “A” on these photographs. Can you find any hazardous conditions that the developer should be aware of?

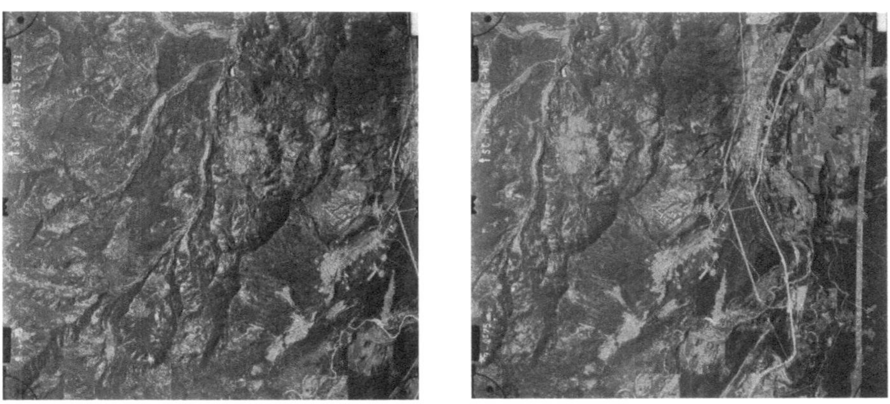

Cortez, Nevada. This stereo pair shows a mountain block near Cortez, Nevada. Determine the tectonic activity of this mountain front. Defend your classification.

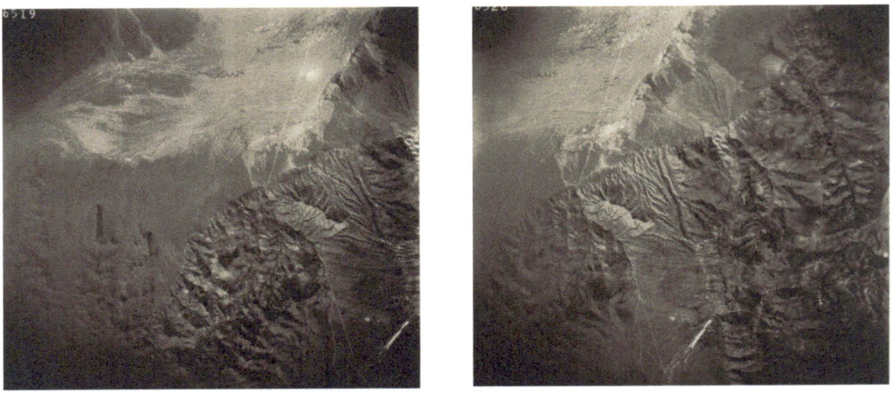

Kolibri 360 GmbH
Ferdinand-Jühlke-Straße 7
99095 Erfurt, Deutschland
produktsicherheit@kolibri360.de

Interpreting Aerial Photographs to Identify Natural Hazards